WASTED IN ENGINEERING:

STORY OF INDIA'S YOUTH

notionpress.com

WASTED

IN ENGINEERING

Story of India's Youth

"From one of India's million wasted engineers"

A book by
Prabhu Swaminathan

notionpress.com

Notion Press

5 Muthu Kalathy Street, Triplicane,

Chennai - 600 005

First Published by Notion Press 2014

Image Source: www.ImagesBazaar.com

ISBN: 978-93-84391-62-1

Dedicated to my parents, relatives and teachers who messed my life up by making me study engineering.

"Everybody is a genius. But if you judge a fish by its ability to climb a tree, it will live its whole life believing that it is stupid."

– quote attributed to Albert Einstein.

If Einstein was born in India, his parents would have made him study mechanical engineering and he would have become a banker after working for a software coding company.

This book is not a work of fiction. If any characters or incidents resemble your real life, then it's purely intentional.

Contents

Introduction: What this Book is About

'Engineering padicha nalla future – If you study engineering, you will have a good future.' This is a claim often repeated to children and teenagers by parents and teachers in many parts of India. But those who have pursued and are pursuing engineering know that it's not completely true.

My engineering education and career have been completely useless to me. There is no polite, polished or diplomatic way of saying this: Graduating from one of the best institutions in my state with the most popular degree in India has not benefitted me in any way! Many of us are either forced or brainwashed to choose engineering after school by our parents, who are labouring under the illusion that it is the gateway to money and success. I always had a doubt; but now, years after graduating, I know for sure that studying what we like and doing what we enjoy may not always lead to money and success but it is definitely the route to happiness.

Some may wonder whether an Engineering degree course deserves this much hatred or attention; after all, it is just another course of study. However, in our time and place, this is indeed a serious issue. We are not talking about just a few random youngsters getting lost in the career highway. We are talking about a huge portion of our population who are growing up without knowing why they are studying the course they are studying and what they want to do with it.

This book is not about any success story. It reflects my failure as an engineering graduate; failures from which you can hopefully learn. Ask yourself why you studied what you studied? What did you want to do in life, when you were in school, before the concept of earning money and going to work entered your mind? Is there any relationship between what you wanted to do, what you studied and what you are doing now? I don't know your answers to those questions, but mine to the last one is an emphatic, "No!" I wanted to do many things in life and engineering was never one of them. I studied electrical engineering and I am now working in a field that is not even connected to any engineering branch. Thanks to the messed up education system and an even more messed up process of educating us, I ended up wasting four years of my life, studying what I did not want to study.

***

Though I am not religious, I always visit temples, churches and mosques whenever I visit a new city or town. They fascinate me because next to engineering education, this is where people invest so much of their energy and money in hope and faith. When I went to Mumbai to take part in the Jagriti Yatra, I visited the Siddhi Vinayak temple (the hotspot of Bollywood celebrities and industrialists). What I saw there not only fascinated me but also shocked me. Beggars outside the temples were getting 50–100 rupees each. In a day, even if a minimum of 10 to 20 people visit them, they get a daily income of 500 to 2000 rupees. And 20 is a low estimate considering that the temple is always crowded with devotees. (Of course, like we see in the movie 'Slumdog Millionaire', beggars do function as an organized group and may not even be able to keep all the money they receive.)

I went there in the 8th semester of my electrical engineering course and I had gotten placed only a few months earlier in a multi-national software company. I was offered a starting salary between 19,000 and 21,000 rupees. (Welcome to the real world! Those 1 crore per year and 1 lakh per month salaries which you see as breaking news are mostly for IITians and sadly, companies that recruit at IITs don't even look at the majority of engineering colleges in India. Forget placements. If you didn't study engineering from an IIT or any of the top 20 institutes in your state, you won't even be seen as an engineer.)

I wasted four years of my life, studying soul-torturing engineering, and faced humiliation by my teachers, just so I can get placed. But beggars here earned as much or more than what I was offered through the grossly over-rated campus placement. Very few realizations can be as humiliating and humbling as that. At that point, nothing else could expose the financial hypocrisy of engineering education better than those begging bowls. They reminded me of how I stood before the interviewer with my resume.

***

I am not an expert on engineering and I don't want to portray myself as one. I am merely recounting my experience. In my opinion, anyone who has spent four years of their life in an engineering college is entitled to voice their honest thoughts about it. Each chapter is independent from the others and looks at several different issues. I don't have any greatly esteemed IIM or IIT degree to add to the credibility of my views but I do hope that the arguments I have made in this book against the most popular educational option in India is seen within its merits and de-merits. I am not sure whether this is coming from a weak academic ego. In this country, any point made with the IIT or IIM tag, is blown out of proportion while most engineers are actually non-IITians. Because of which, their struggle as engineering students is often ignored.

***

We are a nation of misfits. Students of engineering often end up as artists while several arts students study arts only because they couldn't qualify for or afford an engineering degree. There are some who take engineering because they actually like engineering and want to make a career out of it, but these are only a minority among the lakhs of engineering students. The majority of us join engineering with no interest or purpose.

***

In countries that win Nobel prizes in science and technology, year after year, children study what they love and naturally, they are able to create Apples, Microsofts and Facebooks. However, in India, children study what their parents want them to study, which is invariably the course that brings the most money. Then we proudly share on facebook and twitter that 20% of NASA scientists and 29% of Microsoft engineers are Indians. We are not able to create the next Facebook or Apple because we study engineering not to become engineers but to earn money. Money may be functionally important but it is never more important than happiness. We sacrifice happiness for money and then we spend that money to buy back our lost happiness.

***

Studying engineering is looked upon as a high responsibility and a financial compulsion. When we really look into the reasons behind students in India opting for engineering, what they go through in an engineering college and their career path after earning an engineering degree, we realize that this is the most hyped education option in India.

The concepts of few of the chapters might be known to you or already experienced by you. The intention of this book is to tell you that in many cases, studying engineering is a total waste of time, energy and resources. By the time you turn the last page of this book, I hope to have helped you realize the less popular side of the most popular course of education in India. I want the book to give you a reflection of what you have done, what you could have done and what you still can do. My life has already been affected so much by people preaching to me about what I should do and should not do. Now the last thing I want to do is preach others with a book. I hope this book works as a sort of self-help guide on coming out of engineering. Irrespective of whether you agree with me or not, I want you to find an escape route that keeps you happy.

This book can be grouped into two 'I's: Ideas and Issues — ideas about how to make the best use of the worst education option for many and issues that make engineering to be the worst option for most of us.

***

In April 2012, I wrote a short Facebook note on how messed up our engineering education system is and posted it on the day the results for engineering entrance examination were to be announced and signed it as 'Another Wasted Engineer'. It was received well by readers and many were able to relate to it. Encouraged by Facebook likes and shares, I initially thought of converting the article into a story format that includes characters from both engineering and arts background, set within a college theme. And of course, what would any college story be without a little bit of romance, action and fun? In my efforts, I went up to registering the domain name WastedEngineers, but soon I realized that many writers have already published novels that are set in an engineering college campus, including Chetan Bhagat who has also written a successful novel about engineering deviants. There was a risk of being labeled an imitation (the ultimate fear of most writers), so I decided that since the message of the story is more important than the story itself, I should rather write down my thoughts and experiences in the form of a guidebook for frustrated engineering students and graduates like me, using real-life examples of my school, college and Facebook friends.

When I was in college, I nearly got suspended for writing an article in the annual magazine about the lavish amounts of money being used to construct places of worship and stating that it could be better spent on the construction of schools and hospitals instead. I have always been bluntly open about my opinions. If my parents or teachers are reading this, I would like to tell you that the lakhs of rupees wasted in making me an engineering graduate could have been used to open two schools for poor children or atleast a public toilet.

***

This is my first attempt at authoring a book but I have definitely tried harder than what I tried during my semester exams. During the course of writing this book, some people advised me to keep the book 'silly', to make the book attractive to more readers. I decided against it. I feel that studying engineering without any interest and doing a job that you hate is not a silly little issue that can be laughed away. It is a serious issue that is affecting the lives of many Indian youngsters like me. I am certain that there will be many parents who will, after reading this book, declare that anyone asking students not to join engineering to make money is stupid.

To them, I say, India being super-happy is more important than India being a super-power, and one of the many obstacles in achieving a happy India is our obsession with engineering education.

Schools: Science vs Commerce Streams

For many of us, our problems with engineering education begin before joining an engineering college. For few it even begins before they are born, when their parents make plans on how to educate their children even before they mate and reproduce the child. However, for most of us, the problem begins in school, where instead of being seen as a school student, we are seen as future engineering material.

When a school student performs consistently well in history, geography, mathematics, science and language papers, a broad-minded educator would try to analyze where the child's interest lies and try to develop that child's natural life interest. In India however, if you score above 80% in 9th and 10th standard exams, your parents, teachers and school principal automatically decide that you should join a science stream for your 11th and 12th standard.

In many schools, students who score a high overall total are automatically placed in a science stream, and are discouraged from opting for a commerce or literature stream, even if their interest and potential lies in either of those. On the other hand, students who score an average or below average overall total are not allowed to take a science stream because two years down the line, the school administration will not be able to show a better result in their annual 12th standard students mark list. Such students end up in commerce streams even if they had scientific potential and wanted to be engineers or doctors.

***

In your 10th board exams, if you score 99% in language, history, geography and 50% in science, you will still get a science stream for your 11th and 12th standard with the help of your high overall total, even though you may not have any interest in science or have any scientific aptitude. Whereas a student who scores 50% in language, history, geography and 100% in science, has a relatively poor chance of getting into science stream for their 11th and 12th standard, even though his/her scientific aptitude is clearly reflected in the exam marks. In schools, the filter used to segregate the 'studious' from the 'non-studious' after the tenth board exam results is the overall total and not subject-wise performance or life interest. However, once you finish your 12th standard board exams, your science marks matter more than your other subject marks. They even matter more than all the marks you've scored since your kindergarten if you want to pursue engineering or medicine. Whereas, that 10th standard student who scored 100% in science and

Prabhu Swaminathan 50% in other subjects would not even be admitted into science stream.

***

For many schools, the focus of educating children seems to be basically making them into engineers and doctors. How many schools would be ready to say with pride that their topper joined a history or geology course? The pride of the school is built on how many children are made to be future engineers and doctors, rather than how many children turn out to be happy individuals of the society.

It simply does not occur to school teachers and principals that a school topper may also want to study B.A. History or B.A. Economics; a student who wants to study B.Sc. Physics may also study as well as a student who wants to take up electronic engineering.

If you score well in school board exams and if you wish to study B.A. Economics or Political Science, you will be initially criticized, discouraged and demotivated; and later convinced or forced to prepare for engineering entrance exams. But ironically after engineering if you get selected for a post-graduate financial course in London School of Economics, you will be encouraged and praised. The same school that discouraged you before will now invite you to address their students.

If a Masters degree in finance or financial management is the student's core interest, it would make more sense to allow them to take B.A. Commerce or B.A. Economics and then opt for MBA in Finance.

This way, their knowledge in their field would have been benefitted, while more seats at IIT and other top engineering institutes could have been used for others who are actually interested in engineering.

After 10th standard, classification into science and commerce sections works in effect like a separation between the studious and non-studious, which is viewed as two future groups of engineering/doctor aspirants and arts college aspirants. The student who is put in a science stream is severely demotivated from pursuing an Arts degree after finishing school and is pressured to join an engineering college. Whereas the students who studied commerce cannot pursue a science degree even if they develop an interest in science later.

Sports Entry in Engineering Colleges

There are separate seats allocated in engineering colleges for meritorious candidates who have contributed to the country in the sports arena, at national and international level. The basic logic behind this is that these students spend so much time at excelling in their sport because of which they don't have enough time to prepare for the tiring school board exams or the mentally intensive engineering entrance exams. I am completely in favour of the allocation of such seats to those who make the country proud by winning us medals at various games and I do not stand for the removal of such seat allocation.

However, if we look at this through the prism of education, it should make one wonder why a person who wants to make a particular sport their career is made to join a completely irrelevant field like engineering.

This might be because engineering is seen as a back-up option for many of our young sports achievers since sports, especially sports other than cricket, don't offer good career prospects. Many popular engineering colleges have excellent sports facilities and few colleges in south India had once even offered free admissions to medal winners. In some engineering colleges, you can even find well-developed sports facility existing side-by-side with poorly maintained science labs.

If the ultimate aim is to secure the future of the child, it would make more sense if financial assistance could be given to medal winning school students to enable them to pursue a sports-related degree of their choice, in India or abroad, instead of wasting their precious college years while also saving the few seats at top institutions for students with a genuine inclination toward engineering.

***

There was once a popular swimmer who had broken records even before he finished school. He was given admission into an engineering college via sports quota and graduated as an engineer. This was the man who had said in several interviews that he dreamt of making swimming a popular sport in India. Now, however, he works as an engineer. This is one of those classic cases where the sports achievement in school level does not translate into a long-term career. If he was instead offered a job in sports administration or recruited as swimming coach, it would have been a far better way to utilize his talent and nurture his ambition.

On the other hand, there are several school students who perform well and win medals at International Science Olympiads or have great scientific achievements while they are still in school. Ironically, these students don't have any special quota in engineering colleges. Ideally, a school student who is able to publish scientific papers in international journals or demonstrate his scientific potential by making small inventions at a young age should be favoured for admission into top engineering colleges. However, even these kids cannot get into good colleges if they don't have the ability to memorize large amounts of data for their board exams and entrance tests, whereas someone who wins a medal at a sport has an easier chance of getting into a top engineering college, even though their accomplishment is completely unrelated to the field of science or engineering.

***

If you happen to be studying engineering through the sports quota, do understand that this is not meant to undermine your accomplishment in any way but it is undeniable that a vast majority of students take up engineering through sports quota for reasons such as career stability as a sports career in India may not have a lot of scope and definitely not out of interest in engineering itself. The question is why do you join an engineering college when your real interest lies in a particular sport?

For the sake of our convenience, let's assume a hypothetical Arvind, a 12th standard school student, who has won a medal in a national hockey tournament and wants to pursue hockey professionally as a career and would even like to get involved in sports administration or training and coaching after his retirement. The ideal way to support his dreams would be to enroll him in a full-time sports program, possibly with a monthly stipend, and allow him to continue his education through distance or correspondence mode. Alternatively, he can be given a scholarship for a degree in Bachelors of Physical Education in a top college, maybe even an employment offer with the state hockey team or the state sports development board after he graduates. Both methods would give Arvind moral and more importantly financial support to fulfill his dream. This would also encourage other aspiring sportspersons to pursue their dreams. Instead, Arvind gets a seat in a leading institution for engineering thanks to his sport abilities, he ends up graduating as an engineer, unable to give enough time to his sport, he does his MBA and lives the rest of his life working a desk job, never having fulfilled the dreams of his youth.

Youngsters are afraid to consider sports as a career because it does not offer much financial security or a long-term employment opportunity. Instead of giving these youngsters opportunities in a field they might not even be interested in, the government should start a career-grooming process for these young sports achievers when they finish school itself. These players can be given the opportunity to join state and national sports teams or even get into the educational system by coaching for schools and colleges. They can also be given government jobs as sports assistants and administrators in the state sports development boards. If we could create a system that allows them to hold on to their sports dreams, it will contribute to the development of sports (other than the much loved cricket) and more importantly enable them to have a career in the field they love, rather than looking to the seats allocated for them in engineering colleges to provide them with a guarantee of employment.

***

There is a striking similarity between how parents force school children to attend sports training camps irrespective of whether they are interested or not and how these same parents later force their children into engineering even if they want to pursue a career in sports.

In a shelf in my house, there lies an old cup I won in a city-level chess competition. I 'used' to be a chess player in school and I attended most of the chess competitions held in my state. My father and teachers had initially encouraged me hoping that if I could get a national or a state medal, then getting into a top engineering college through the sports quota will be very easy.

When it became clear in my 11th standard that my sports talent was not good enough to win a medal, they decided for me that I should put away my chess board and open the board exam books. The king was check-mated. Soon, I went from contemplating chess moves to constantly flipping through guides for my entrance test. After finally getting into an engineering college, I tried to play chess again but I couldn't. As any sportsperson would testify, it is not easy to resume a sport after a long gap. The cup has since gathered dust and it's been a while since I even thought of cleaning it.

The Sports Authority of India or the respective state's Sports Development Boards could start a separate sports educational institution with syllabus and curriculum designed for the young sports achievers, with an employment offer after graduation, in the sports administration field. When we can spend crores of rupees in Indian Institute of Technology and Indian Institute of Management, why not build and promote Indian Institute of Sports along those lines? The best way to respect sportspersons is to give them a career in sports.

***

Many engineering students, after graduating and working in an I.T. company tend to deviate into fields like photography, cinema and business management after a while. Some also succeed nationally and internationally in their chosen field, however, for a sportsperson this is much more difficult. For them, age is a very crucial factor and wasting four years of their youth in an engineering college may cost them their entire dream.

Engineering is very stressful and it sucks away their energy. It would be difficult for an aspiring sportsperson to stay emotionally focused on their sport while the academic pressure causes such immense mental strain. When the engineering semester exam dates clash with their sport selection trials, the trials are obviously skipped in favour of the exams. Soon scoring goals becomes less important than clearing semester papers and getting placed in their campus interviews becomes more important than getting a place in the national or state team. During their college life, it becomes clear to them that there is no monthly income opportunity in sports and they then realize that getting a job as an engineer is more 'practical' than getting an employment opportunity using sports.

The fact that joining an engineering college is a popular option even among medal-winning sports achievers shows why our medal tallies in the Olympic score board are at the bottom.

Factors at Play: What makes Someone join an Engineering College

Who is the real decision-maker?

Parent-child relationship in India is often a one-way street. Though things are changing a little, in most parts of India parents are still treated as demi-Gods and children are expected to obey parental directives without asking any questions. This authoritative parenting style affects every personal and career decision of an individual from his/her childhood to adulthood.

When a child is born, it is natural for everyone to expect the parents to take care of the child completely, and ensure that the child's nutritional, educational and personal requirements are taken care of. This stage is controlled by the parents but when the child enters teenage, the parents should know that the 'taking- care' phase is starting to fade out. The child who has reached teenage is now learning slowly to take care of their increasing personal and social needs.

But most parents continue to see their child as a child throughout his/her life. While this is a remarkable testimony of parental love, it can become suffocating for the growing youngster. Often parents end up making the decisions for their children long past the right age. Many parents don't know when to let go of their child and when to stand by their child and guide them.

Choosing the right college and the course of study after school is an important decision for an individual. And it is natural for a person to consult their parents while making this decision. Sadly, in India, parents tend to use this as an opportunity to live out their dreams by dumping all their dreams, plans and ambitions on the child. All too often, what parents want the child to study in college becomes more important than what the child wants to do in life.

Schooling:

If you top all the exams or even study moderately well in school the general assumption made about you is that you have already decided to join engineering and you are studying so hard only to get an admission into the best engineering college. The thought that a student who wants to study B.A. History or B.Sc. Physics can also study as good as a student who is preparing for engineering and medicine does not occur to many, even to school teachers and principals. Now the general trend is that whoever studies science stream in school is 'encouraged' to join engineering.

General economic condition of the family:

Not everyone who wishes to study engineering becomes an engineer and not everyone who studies engineering wants to become an engineer. Money unfortunately plays a crucial and cruel role in preventing many children who want to be engineers from joining an engineering college. Money does not affect them at the college admission entry alone. The damage starts much earlier. In all probability, a child from a low income family would have studied in a school that is quality-wise poor, which would have in turn impacted their knowledge and performance. This will inevitably diminish their chances of getting into a good university/college later.

Some poor parents join their children in an engineering college after getting expensive loans with the hope that their children will be well-placed after they become engineers. If the student has an interest in engineering or at least sees the engineering degree as a gateway out of poverty, then the student can make a decent attempt at getting out of college with a job in hand. But if the only reason a student is pursuing engineering is parental pressure ('our family now depends on you' and 'you have to study engineering to earn money and take care of us'), chances are there are plenty of tears down that road for all concerned.

EEE: Engineering, engineering everywhere:

Though every child is unique, children tend to imitate the social pattern around them as they grow. They follow what people around them do. They learn from those who grow up before them. When everyone around them talks about engineering, when they see their school seniors join engineering colleges year after year, when they see their neighbours and cousins join engineering colleges, when they see uncles and aunties around them talk with 'pride' about their sons and daughters in engineering colleges, the children tend to assume that joining an engineering college is the natural route after school. Many realize very late in life about the various popular arts courses they could have taken instead of engineering after school.

Brand value of the college: College over Course

Among the various factors that influence the choice of under-graduate education, the choice of college often takes priority over the choice of course. In other words, the engineering college one joins is given more priority than the engineering course one takes.

If you had just finished your 12th board exams, the question you will be asked often is 'Which engineering college will you be getting for your marks?' The brand value and popularity of the college overrides the courses which the college is offering and this brand value is not linked to the quality of education that the college offers but around the placement results.

Even if there are no teachers in a college with 100% placements, it will still be seen as a great college to join. On the other hand, even if a college has Harvard and

Stanford professors but no placements, then it will not be on the choice-list of many parents. Getting a job after finishing engineering is seen as more important than studying engineering.

I.I.T. Brainwashing

(Brain washing may sound like a harsh word to describe the process of joining I.I.T. but considering the young age at which the children are made to join coaching institutes, it's hardly an overstatement.) The Indian Institutes of Technology popularly called as I.I.T.'s symbolize some of the best Indian talents in engineering studies. It is hailed nationally for producing world-class engineers even though the I.I.Ts themselves rarely figure in any global top list of engineering colleges.

Getting into an I.I.T itself is seen as a tremendous achievement. Once you graduate from the institution, you have the life-long brand value of being an IITian, which you can also fruitfully use for gaining admission into a management degree at an Indian Institute of Management (I.I.M); even though it is a totally irrelevant field when compared with the 4-5 years of hard-core engineering subjects which some of the best teenage minds of India study in the I.I.T system.

Within India and according to the coaching centre advertisements, the I.I.Ts are the best educational institutions for those finishing school and it is the responsibility of every parent to register their 13-15 year old kid in an I.I.T coaching program. Ironically 'programming' is the correct word to describe how the children are taught to memorise and vomit, even before these 13-15 year old kids learn to think on their own.

Despite the global projection of I.I.T as a byword for quality education, the entire process of preparation for entrance exam, selection and choices of courses at various I.I.Ts all over India is a classic example of how our education system destroys our children.

- Preparation for I.I.T entrance exam begins at 13-14 years of age, normally when the children are in their 9th and 10th standard. Can a kid who is just 13 years old take the decision of studying in an elite engineering college on his own? No. It is the parents who make this decision.

- The student who wants to study chemical engineering takes up electrical engineering just because he/she gets a seat at an I.I.T. Even among I.I.Ts, computer science engineering in an old I.I.T. is seen as better than mechanical engineering in a new I.I.T. These choices are made with no reference to the student's area of interest. This forced deviation of choice from something you like to something which society likes is relatable to most of us, for the same society has molded and shaped us.

The parents take more pride in saying their child is studying a random course at an I.I.T than a course the child likes at a less popular college. This is also actively encouraged by coaching institutes and schools, whose brand value will rise when they showcase your passport size photo as their I.I.T. product in their full page advertisements in the front page of national newspapers. Promptly, both the fees of those coaching institutes and the schools that produced I.I.T. joiners will rise, further pushing out children from poor economic background from accessing the coaching facility.

- This point might not be a valid argument but I cannot resist noting the fact that almost all the famous I.I.T. graduates in India are artists, bankers, social activists, writers and marketing managers.

***

Many IITians don't understand what is wrong with the IIT system. The problem does not lie with shortage of professors and lack of high-tech labs. It lies in the input. When school students are forced and blackmailed to join IIT, we should not expect the engineering students in India to perform better than their counter-parts in MIT and Stanford where students who love engineering take up engineering education unlike here where we join engineering colleges only for a placement in an I.T. company and then for a banking career. Normally you would expect scientists in the famous alumni list of an institution of science and engineering. Ironically, but understandably, in India, we see bankers and stockbrokers in the popular pass-outs of an engineering institution.

It is funny how so many IITians protest over how the seats should be filled and what entrance exams should be conducted, to allow only the brightest of kids to enter the great halls of IIT when it is clear that pushing students to join engineering affects them psychologically, and the results dog them all their lives. That is the biggest 'quality' problem for I.I.T. You don't need an IIT or IIM degree to figure that one out.

Girls in Engineering

One of the few positive social changes that have come out of the race for engineering degrees is the increase in the number of Indian parents who are willing and ready to send their girl children to study away from home in hostels. This section of parents constitutes a minority but a growing minority who should be encouraged further.

Women, like everyone else of any gender, should be free to make their life choices. Unfortunately, this is not the case in our country that professes historical worship and divine respect for women. The best way to respect women is to let them do what they wish to do with their lives or at least not stand in their way.

Often the girl students in India are forced to make choices based on factors that are gender-specific and based on outdated conservative notions. This includes the choice of what to study and where to study.

Distance from Home:

Parents who don't want their girls to travel far from their home choose near-by colleges even if the fees are high or the standard is low. These girls might end up studying in a nearby Arts college when their interest could be in an engineering college faraway or they study in an engineering college nearby when their interest is in an Arts college which is faraway.

Such controlling parents tend to always keep their female children closer to them even if they are travelling out as a family group. When these girls join an engineering college, they miss out on the Industrial Visits to other states and project visits that require staying outside their homes.

Some rich parents use their money to destroy all forms of predicted and expected social interactions. I remember one of my college-mates once had to travel outside the city for her project work. Her father paid additional charges and brought the equipment straight to her home. He knew that she will be travelling only with her female classmates but he didn't want her interacting with the project support centre staff.

Though we have made so much social progress nationally from the days of stopping a girl's education right after school, much more remains to be done to make parents think rationally about the education of their daughters.

'Strict' Colleges:

Colleges run by social conservative groups that are infamous for separating boys and girls everywhere from buses to classrooms to canteens are a favourite among parents who want to isolate girls from any form of social interaction with boys. Such colleges market these tactics as 'discipline'.

Traditionally, arts colleges are known for their liberal campus atmosphere which doesn't go well with conservative parents who find the engineering colleges that are known for punishing campus lovers 'comfortable' and 'safer'. These parents want the girls to return home from college just as they were when they left home. The only addition that they want to happen to a girl's mind is 'educational' input and what they see as education refers to only nuts, bolts and codes. They don't want the girls to receive the social education that comes from interaction with people.

Many parents can be seen standing next to girls while they wait for the college bus to pick them up and when the bus returns late evening they return to the spot to check whether their girl has not skipped the class.

Dowry Savings:

Many girls who want to be doctors, scientists and engineers are not able to take up their desired course because their parents feel it is better to save the money for her future dowry to get her married off.

These parents don't hesitate to spend 5 lakhs to rent a grand marriage hall to impress the bride groom but don't wish to spend two lakhs for her engineering college fees.

Preference for Boys:

Consider a conservative middle-class family with a girl child and a male child, who are finishing school education at the same time. It is most likely that the girl would be groomed for a commerce course in college, even if she wants to pursue engineering and the boy will be the chosen one to pursue an engineering degree even if he has no interest in it.

Some parents crush the dream at 10th standard itself and make the girls join a commerce/economics course, so there will not be any chance for the girl to join engineering or medicine after she finishes her 12th. People discourage a studious boy from joining a commerce course after his 10th standard, but it is not frowned upon if a studious girl with interest in science does the same. Neither the school nor anyone else motivate the parents of the girl to enroll her in a science course. The 'interest' which schools show on the education and college plans of boys is not shown on the college plans of girls.

Even when these girls ultimately join an engineering college, such parents tend to marry them off as soon as they graduate or even before they cross the age of 25.

Soft Courses:

If parents decide what branch of engineering the girl should study, then 'soft courses' like computer science and information technology have more chances of being opted than mechanical and automobile engineering.

Mechanical and Automobile courses are seen as 'rough' and any course that does not involve field visit and travel is considered better by these conservative parents.

***

Engineering Housewives: Obviously, most of the toppers in my college were girls but sadly many of them have become housewives after marriage and have quit their career permanently after having children. Since I know them personally, I assume hopefully that the decision to quit their career and raise children came from a personal choice and not forced by their husbands. When I asked them what they wanted to study in college, lot of them had a non-engineering idea in their minds, from history to horticulture to public relations. Their parents had made up their minds when these girls were in school itself that they should get married before they were 25 years of age and that they should quit their career if their husband wants them to. The least these parents could have done is let these girls study what they liked, instead of framing their marriage plans and choosing their husbands, even before they graduate.

There are also some men who want only an engineering graduate for a wife. They feel it is below their standard to have a non-engineer as a wife and at the same time, they want her to forget her career and take care of her home.

Infamous Qualities of Notorious Engineering Colleges

There are various factors that contribute to a miserable college experience — non-understanding parents, being unable to develop or carry out a relationship; overly conservative college authorities; inefficient teaching faculty; aggressive non-teaching staff and genuine lack of focus or the belief that you lack focus in life, future and career.

***

There are arts-college type engineering colleges that don't care about what you do. They don't care whether you study, whether you attend the class or whether you even pass the exam. You are free to do what you want to do. There are liberal colleges that pay attention to your marks and exam performance but don't otherwise bother about what you do in the classroom or campus. And there are the highly conservative, what we call Hitler-type colleges, many of which reflect a rule of dictatorship. These colleges control every aspect of a student's life. Not a single action of yours will escape their eyes. Any minor deviation from the standard practice established by the college management invites punishment.

It is these Hitler-type colleges that are the worst category for all students irrespective of whether you fall under the 'studious' or 'non-studious' category. I include 'studious' students also because in other colleges, if you study well, the college staff will not harass you with rules. But in these colleges, even the academically well-performing students are harassed with 'rules'.

Sadly, lot of our engineering colleges fall under this type. A few of their many ways of harassing their unfortunate students are:

Internals:

About 80% of our overall semester marks come from the subject paper we write in the final exam and 20% from the internal assessment of the department head and class teacher. It is unclear both at national and state-level whether there is any standard procedure as to how this internal assessment has to be done. While some colleges are liberal and give above 15 marks out of 20, keeping in mind the interest of the student, many colleges use a combination of attendance, class tests, mid-semester exams and 'behaviour'. There is no standard or pre-fixed procedure.

In colleges where the teachers, heads of department and the principal decide the evaluation procedure, internal assessment does not serve the purpose of assessing the students but often serves as a blackmailing tool to break the backbone of students who 'rebel' and don't fall in-line with the college management.

Leave applications:

In most universities, 75% to 80% attendance is the required percentage that will enable a student to appear for semester exams. A semester is for six months, which has about 180 days. If you cancel out the weekends, national holidays and final exams, we have about 150 working days. According to this rough estimate, a student is eligible for at least 10 to 20 days leave in a semester. However, most engineering college students will know that it is very difficult to obtain even a few days leave for any important work. If 75% attendance is necessary or enough, then the student is automatically eligible to take the remaining 25% of the working days off for their personal purposes or in case of an emergency. This reasoning, however, is not accepted by the colleges.

The issue of attendance becomes even more difficult in colleges where the internal assessment is linked to the attendance. Here, a student with 100% attendance will have more chances of scoring a full 20 in their internals than a student who has 80 or 90% attendance. Even if the students maintain the required attendance, they stand chances of losing out on their internal marks.

There is, of course, a great exception called the 'on-duty' leave, popularly known as 'O.D'. This is granted for missing college due to an educational event you have taken part in. But their graciousness ends there, as their obsession with rules resurfaces again. O.Ds are only granted if you appear for an event related to engineering, not if you wish to participate in a literary, artistic or social event.

Dress code:

No matter what the moral guardians of the society say, students do know the difference between an educational institution and a sea-side beach club and can dress appropriately according to the occasion. However, colleges have made it a part of their educational agenda to decide how the students should dress. The most affected by these dress codes are the girls. They are brain-washed, from a young age, into thinking that they have to dress in a certain way to be respected and to protect their culture. If they fail to do so, they are told that they disrespect and hurt our tradition.

Almost in all engineering colleges, the mandatory attire for girls is a salwar-khameez or churidhar. Several conservative colleges even have 'dress squads' to check whether the imposed rules are being followed.

On cultural occasions, girls are asked to come dressed in sarees to promote the Indian culture. I've always wondered why when we boys can come dressed in shirts and pants even for cultural celebrations, which have nothing to do with Indian attire or tradition, why not let the girls dress in shirts and pants as well? It is ironic how shirts and pants become Indian wear for boys but become western wear for girls. This again shows how we expect women from a young age to carry the burden of culture, while men can dress and behave as per their wish and comfort.

How can a supposedly scientific educational institution send a female student back home for wearing a shirt instead of a salwar? It's a shame that in India, we have engineering institutions that judge a woman's scientific capacity based on her dress instead of her intelligence.

Many colleges and universities have even completely banned jeans. They seem to associate jeans with indiscipline and a careless attitude. Mark Zuckerberg and Steve Jobs wear jeans at work and have created life-changing technological products. Here, colleges that ban jeans under the excuse that students will study better differently dressed cannot even guarantee 50 percent campus placement.

Dress code is another area where engineering colleges can learn from Arts colleges. Most arts colleges don't impose dress regulations on the students; neither do the students misuse this freedom. In popular arts colleges, you will find that their dressing shows a lot of variety within accepted limits of decency and public decorum.

Moral policing and invasion of the personal rights of students:

Educational institutions, especially those in south India, think that just because a student studies with them the personal and social life of the student must be regulated by the college, in the 'interest' and 'welfare' of the student. Many top engineering colleges and universities in India have their own definition of 'engineering discipline'. According to them college is a place to only study and socializing is not allowed. Many colleges run by social conservative groups have taken extreme measures to prevent boys and girls from talking to each other. These colleges also have separate seating arrangements for boys and girls from classrooms to canteens. All forms of social entertainment are strictly prohibited and the focus is kept only on classroom-based education; which, by the way, is often not great in such colleges.

One of the inside jokes we had in college portrays this trait of engineering colleges very effectively —'Welcome to Tamil Nadu's engineering colleges, where HODs fail you if you are caught speaking with a girl.'

Certificate Blackmail:

Many students would have experienced this at school level itself. Schools and colleges generally collect the birth, transfer and educational certificates from students which they only return after you complete the course. Consequently, if you get into any trouble with the management and want to get transferred to another college, you may not be allowed to until you pay the full or partial fee for the remaining years.

***

Patience — the golden virtue required to survive in such engineering colleges.

Imagine your parents being forced to travel a hundred kilometers to visit the college because you questioned a college rule, because you didn't give 'enough respect' to a teacher or because you failed in a class test.

What you just imagined is an everyday reality in several engineering colleges in our country. All these rules are painted as discipline and its violation is treated as 'indiscipline', which is where your control over yourself comes into play. It is best for you not to lose your temper before the teachers or principal, under such moments of injustice and humiliation.

There is no national definition for 'indiscipline' within the context of educational institutions in India. Colleges and universities have framed their own definition to fit their dictatorial, misogynistic and totalitarian attitude. In one college talking to girls is considered as indiscipline. In another, failing in exam is indiscipline. While colleges usually don't selectively pick and target individual students, when the students happen to break these college rules, they are marked out. There are teachers and heads of departments who deliberately fail and punish students who may have questioned the rules of the college or for supposedly failing to respect people in authority sufficiently.

Colleges have excessive control over students' life, from their internal assessment marks to their social behavior. If you have joined such a college, you are very unfortunate and deserve sympathy, but you have to push yourself safely through the four years. It is also advisable to keep a mental track of the few liberal teachers.

***

Future interaction with college:

If a student intends to go abroad for higher studies, they are required to have a 'character certificate' from the college they studied in. Unfortunately, in our educational system, character is judged based on how many rules you break rather than how many rules you live by. Another terrible way that colleges use to assess the character of their students is based on their mark sheets. As if a student who scores 90% has higher ethical and moral values than a student who has arrears.

It is advisable to avoid any intense confrontations with the staff or principal of the college if you have a Masters course planned for the future. (Perhaps western and European colleges should understand that in Indian engineering colleges, you have a 'good character' only if you keep your mouth shut and follow the barbaric rules.) The brighter side, however, is that a lot of colleges often give a standard character certificate to all students.

***

Six ugly incidents that happened in engineering colleges in and around Chennai:

1) Ramesh was studying in a popular engineering college near Chennai. On his classmate's birthday, he shook her hand and wished her while they were waiting for the college bus to come, miles away from their college campus. The college bus driver saw this and complained to the college management who immediately suspended the boy for 30 days. His suspension led to a deficiency in attendance and he was hence unable to appear for that semester's final exam.

2) During a random college raid for cell phones, a student named Abhishek had his cell phone seized by the college Physical Trainer. (Only in India, we find sports coaches checking for I.D. cards and cell phones, instead of teaching physical exercise.) While going through Abhishek's messages in the inbox, the P.T. coach found a joke from a girl named Ramya. He had saved that girl's name along with the course she was doing as 'Ramya ECE'.The coach called the number to confirm it was the same girl and when she answered, both students were immediately taken to the principal's office by the P.T. coach like a prized loot catch from a war. Fortunately for both of them, the matter ended with a fine of 10,000 rupees each, rather than a blow to their degree. Assuming that carrying a cell-phone is against college-rules, even though it is not against Indian laws, it still doesn't

give the college authorities the right to invade the privacy of the student by reading their messages.

The students could have gone to court but decided not to, in spite of facing a lot of humiliation, because they knew that a suspension could have made things worse and just paying the fine, however large the amount, is the safest way out, since future internal marks and co-operation during practical exams are in the hands of the college administration, where they can extract revenge, if the students did take the matter to court.

3) Vikram, another engineering student in another college, got into an argument with the college authorities after the college management had cancelled the permission given for a cultural event. After refusing to apologize, he was suspended for two weeks. He filed a court case against the college management stating that arguing is not an act of indiscipline and the court ruled in his favour and asked the college to revoke the suspension. Before filing the case, however, he failed to take into consideration his own academic position within the college. His attendance, being already low, the college used it as an excuse to prevent him from appearing for the semester exam. This time, he could not file a court case, as the college followed the university rules. He then wrote all the papers in the next semester as arrears and also missed attending the campus recruitments.

4) "Girls at 17 can be easily influenced. They need the injection of discipline and I give them that. We are equally strict with boys as well. Students come to study, so why they should talk with each other? Do people talk to each other in temples and churches, so what is the need to mingle in colleges? " - This is a statement made by a University Chancellor and owner of many educational institutions on boys and girls interacting with each other in colleges. Now you know why so many of our youngsters grow up without being able to differentiate between flirting, talking and harassing.

5) One of India's leading private universities removed two female students from the college hostel after they questioned the discriminatory rules imposed selectively on female students which stated that the girls can leave the college campus only once a week and should return by 8 p.m. This 'grant of freedom' is given only after their parents send a fax stating their permission. The boys, however, don't need any approval letter from parents and can leave and return to the campus as they like. In a democratic country like ours, girls have all the right to even question the Prime Minister if they feel discriminated against but in an engineering college they get suspended for questioning a college rule.

6) This outrageous incident happened in an engineering college run by an industrial group. The college administration had installed CCTV cameras in student areas to monitor them and the chairman

could watch the CCTV recordings in his room. One day he noticed a girl talking with her classmate after their class got over. Immediately he summoned the girl and scolded her for talking to a boy. As a result of the humiliation she faced and from the fear that the college would inform her parents about the incident, she went to her room and hanged herself to death.

Stating that this last incident was only one of many such instances where an engineering student has taken the drastic step of killing themselves due to humiliation or pressure from the college, I now leave it to you to decide whether these colleges exist to improve the lives of students or to torture them.

***

From frustrated students to bad citizens:

Gender-segregation is a common occurrence in many engineering colleges run by social conservative groups. Imagine the quality of students who graduate without knowing how to talk to the opposite gender, how to say 'hello' to the girl or boy they like or how to even differentiate between flirting and talking. Restrictions imposed by college authorities on the social life of the students often have a bad effect on the society once these students leave the college and become an active part of the society.

One of the best places to promote gender equality is the place where our boys and girls study and one of the best ways to create gender equality is letting girls know how men think and by allowing boys understand what women want.

Instead, what the boys and girls of our society know about the opposite sex is often based on their personal interpretation of novels and movies. We raise boys and girls as two separate species in schools and colleges, and after becoming adults, they struggle to understand the other gender. In 21st century India, several schools and colleges, especially those in south India still prevent boys and girls from talking to each other. I believe that such places of education are also responsible for the existing gender inequality among our people.

***

Violating college rules is not the same as violating the rules of government of India. What is seen as a crime or a mistake under the college's rules and regulations need not be a crime under the Indian Penal Code or the Indian constitution. The law of the land guarantees us certain fundamental rights and freedom which cannot be taken away even by elected governments. Colleges and educational institutions have no right to restrict this freedom given to the student. Unfortunately, many colleges think they are above the law.

What to do if you have Already Joined an Engineering College?

It is always good to make the best out of a bad situation, especially in an unchangeable and irreversible situation like this. Unless you are in your 1st semester and you have secured admission to a degree course of your choice in another college or the same college, it may not be practical to quit engineering altogether. However, if you are certain that you have discovered your life interest in another course, then you could consider quitting even after your first year, preferably after getting career counseling. Otherwise, this journey will have to be endured.

***

There are two kinds of students who hate engineering. First would be the student who is forced to do engineering against his/her interest. And the second is the student who would have joined the course with some interest but over a period of time, the academic and college stress would have destroyed all the interest and liking the student had for the engineering degree.

If you like the course and want to be an engineer, you may not be among the students who face academic hurdles within the college. You can use the four years to improve your technical knowledge and explore your career options related to the field of engineering. However, if you hate engineering and if you want to change track to a non-engineering field after college, then it is better to start planning from your first or second year itself. First, take a month to cry and cry as much as you want for taking engineering. After you have cried enough, wipe your tears away and realize now the best thing you can do is trying to get the maximum benefit out of your life as an engineering student that you have probably started to hate.

Get placed or Try to

It is good to make use of the only reason why many students join engineering colleges — campus placements. Even though you may not be interested in pursuing a career related to engineering or IT, it is financially safe to have a job at hand. Usually there is a one year gap between your final semester exam and your date of joining. You can use this time to plan or prepare for your masters in a course where your interest lies. If your ultimate goal is to start your own business or social venture and if your parents are not financially well-off, you can use the employment to save for your future plans. However, you need to ensure that cares and the responsibilities of an adult life don't trap you in a profession you don't want to be in.

Campus placements are always stressful. You see a wide range of human emotions among your classmates, from anger to betrayal to sympathy. Many start hiding mock placement papers from their friends and some even deliberately misguide friends. Internal politics that have been going on for years come into the open. It is advisable to use this experience to learn more about yourself and others.

Fun tip: If you get placed and reject the job offer, to start a non-engineering career path, it adds glamour to your new venture. You can proudly say that you rejected a job offer at a Bangalore I.T. company to follow your passion.

Serious Tip: If your parents are very conservative and strict, you can use your campus placement as an escape route.

Take up an internship:

There are two types of internships which you should take during college.

i) At a company related to your course of study. This is important because it will help you understand whether your mindset is suitable to work in that field for the rest of your life. If you are pursuing an electronic engineering course, apply for an internship in electronic companies, where you will get hands-on experience in the job profile of an electronic engineer; and you can check whether you are ready to push yourself every day of your life to design machines, draw circuits and develop signal systems without having any interest in it. If possible, talk openly with the company's manager and get a practical understanding on what you will be doing when you work in such a company.

ii) Take an internship in the field of your choice. If your choice of college study and life interest is the same, then you can give yourself a pat on the back with the understanding that very few youngsters in this country find themselves in the same situation. This second type of internship is for the large majority of our youngsters who are not so lucky.

If you want to be a writer, work at a bookstore or a publishing company. If you want to be a photographer, apply for a part-time photography assignment in a local magazine. Whatever your field of interest is, take a list of companies in that field and check for the ones in your city or nearer to your college. Tell them that you are considering a future in their field and you feel an internship there will help you assess your potential. Whatever your choice is, do realize that deviating from engineering is nothing to be ashamed of. Obviously when you shift from engineering to non-engineering field, there will be cynical arguments like, 'you wasted four years in engineering', 'you could have given the engineering seat to someone else', 'you should have taken up arts', etc. It is better to waste the four years in engineering and shift to your field of interest now, instead of regretting when you are in your 40s that you have ended up wasting your entire life in pursuing a career without any interest.

I have myself done internships in both engineering and non-engineering fields. During my second year, I underwent inplant training in one of India's leading telecom companies, where I gained firsthand experience with electrical and electronics stuff (two shiny words in the name of my engineering degree) and realized that my brain just cannot compress itself within a cabin and plan wiring and circuiting. I spent more time looking at the birds outside than at the circuit diagrams in the class. My mind was at its happiest when I worked in a book store and for a child rights group. I wanted to do a lot of things not connected with engineering, like photography, travelling and writing. Many of my interests are in the fields of arts and society. Working for both an engineering and non-engineering company during college itself helped me realize that engineering is not the path that I should take.

Plan your time

Contrary to the popular belief that engineering students are busy all the time, they do have lot of free time at hand, especially the 3-5 hours spent in travel if you are a day scholar. Make use of the free time you have now because once you graduate, you will not have so much of free time ever again. When I look back at the hours wasted everyday in classrooms and labs, I tell myself that I could have used it to improve and prepare for my future ideas. Since I wanted to write, I could have used that free time to improve my grammar and vocabulary. Although it was not a complete waste, I now realize I could have done a lot more.

You can use your free time to gather information about your ideas, thoughts and future plans. For example, if you want to start a social service group, use the time to read about various types of social service and find out which social activity suits you; if you want to be a writer, start improving your grammar and vocabulary; if you want to do a Masters abroad, start checking the available scholarships and the eligibility criteria. Some require a percentage above 80% and most require a positive 'character' certificate from the head of department. Even if you study in the worst college, the time you have after college hours is still in your hands.

Information is the key

Successful people are not the people with money; they are the ones with the information they need and the ability to process the information to achieve their goal. Try to gather all the available information about your college, college's locality and the companies and other educational institutions around your college to get internships or for other future use; keep a note of all the social and cultural clubs, events and activities that go on in your college. You can use them as a testing ground for your talents.

If your interest is in writing, you can make use of the college magazine to practice your writing skills and check the response to your writing; you could also get your photographs published in the symposium and annual magazines.

If your interest is in singing or music, offer your skill during the college cultural events. Don't think that you will get other opportunities after your college graduation to sing or use your music instruments. Performing before a college audience as a college student is one of the best experiences one can have. Overcome the fear of classmates who will laugh at your voice, music or skill. What you will face in the real world industry will be worse. Get used to mockery and praise when you are young, so you don't have to worry about either of them when you are older.

If your interest is in social service, you can make use of the NCC or NSS facility available, to understand how social service works on field.

Yes, you may hate your college but don't hate its resources. Compare your future plans with the resources currently available within your college and try to get maximum benefit out of it. Like the soldiers in the military who do a dry run before launching the actual assault, you could use these resources as a practice or dry run for the 'real life' implementation of your ideas.

Browse your college's website, do some research, check its placement reports and keep a track of alumni events. One of your alumni could be already well established in your field of interest and you can seek their guidance.

***

Avinash wanted to be a professional gamer and he looked at books as a waste of time. Since he didn't want to get placed in any software company, he became careless about his arrears and couldn't clear them before his 7th semester. Much to his surprise, a major digital video gaming company which skipped the previous year's campus recruitment came for campus placement and sadly he couldn't attend because of his pending arrears. If he had taken the effort to check the placement reports of the previous years which was available both in the college library and website, he would have seen the name of the company whose video games he loved and things might have played out differently.

***

Travel as a student

Travel opportunities that are commonly available to engineering students include annual industrial visits, sports and games tournaments, other college cultural events and technical symposiums, and visits related to social activities of the college. Make use of all of them. The more you travel, the more you learn about the world around you. You can make use of these visits to learn about new places, meet like-minded people, 'test' your core interests from singing to robot designing. The best thing about participating in competitions is the participation. Register and attend as many competitions and events as possible.

You have more chances of meeting someone who is interested in music technology as a career in a cultural event than in your classroom. You can meet someone who shares your point of view on gender violence or rural development in a social event than your college campus.

Align your post-graduation plans with your under-graduate life:

Irrespective of whether your post-graduate plans are related to engineering or not, you can make use of your under-graduate college life to learn more about your desired choice of future education, i.e., a Master's degree.

If you are unable to choose a post-graduation degree during your time in college, then don't push yourself into a regular MBA/ME course just because your friends and classmates are doing it. You can register for GRE/GATE/CAT class and attend the exam too but don't join a Masters in Engineering or Business Administration if you have no interest.

Take time to understand where your life interest lies and decide accordingly. Sometimes you may also decide not to go for a post-graduate degree at all, which is also fine, as long as it suits your idea of your future plans. Don't repeat the mistake of following the crowd, like we did while joining engineering after school, like goats being guided to the mutton stall to learn the art of making biryani. Already four years of youth is sacrificed. Two more is unnecessary.

The students are not as confused as they are projected to be. They suffer from a collision of multiple thoughts – all flooding their minds at the same time. Sometimes they want to do this. Sometimes they want to do that. And sometimes they want to try everything. This should not be seen as a negative trait. Some teachers and so-called elders condemn this roaming of mind saying that this indicates your mind and thoughts lack focus and clarity. Only a robot thinks in a straight line. As humans, we are bound to think in many different ways and our thoughts on interests and career plans go in different directions at different times. College days are the best time to experiment and experience the diversity and difference that life offers.

Build the foundation for your after-college plans from college itself. If you want to be a dancer, join a dance coaching academy. If you want to be a singer, start training. When you waste lakhs in an engineering degree, you can surely spend a few thousand to learn professional singing or dancing.

If you want to be a photographer, you can start learning about cameras and photo-techniques, shoot for free at the college symposium and charity events, and pursue a Master's degree in visual communication. If you want to be a social or political activist, you are better off doing a Master's degree in social science colleges instead of engineering. You can even study law after engineering and become a lawyer. This way you can earn as well as spend time for your social service.

Though a huge variety of options are available when it comes to post graduate studies—from electronic journalism to fashion designing—I would like to still insist that a Master's degree is not required for most entrepreneurial and social activism work.

Broadcast your ideas and interest

Find classmates, juniors and seniors who are like-minded and have similar life-interests. You can also connect with more people and contacts through them. Put your Facebook addiction to good use. Check their Facebook profiles, add your classmates and browse what they have added as 'interests' on their profile. You could find someone sitting in the 3rd row with a similar interest in creating music apps or energy production from nuclear fusion.

It is also important to have at least a 'talkable' relationship with everyone in your class or batch. When my friend launched her fashion store, she later came to know that a classmate is good with designing but she hesitated to seek his help since she never spoke to him in college and she then had to hire a designer. Classroom politics often pushes one set of students against another. Though mob behavior and groupism has a human nature, ensure that you are in a position to at least call anyone in your class and seek their help or guidance.

Hate people if you don't like them but don't show it to them. The only useful lesson I learned at weddings is how to act as if we like some relatives even if we hate them. We maintain this 'fake' show of joy at meeting them, (even though our mind is on the food), because we know that we may have to seek their help anytime in future, in case of a family emergency. Same applies to classroom and college relationships too.

Anyone who has created a Facebook page will know the power of human connection. We first invite everyone to our page and then inbox everyone asking them to like and invite their friends to the page. No one is an island. We still need people to spread the word about our work.

Stay focused and Be happy

Keep your focus on your long-term goal, be it doing a post-graduation in your field of interest or pursuing a non-engineering idea. Keeping your focus on the life that awaits you after the four years of college will help you cross the engineering college life with determination and comfort. It helped me take the stress away from the four years of engineering college life. I found happiness in college by planning what I would do after college.

Some advice that after you finish engineering good things will happen to you. They will not happen on their own. While optimism and a positive outlook towards life is important, don't let them blind your eyes. We have to build our own happiness. There will be students in your college/class that would love engineering and have a scientific bent of mind. Don't feel guilty or feel bad about yourself. It is not your fault if you don't love the subject you are studying. Learn from them. They are doing and enjoying what they like. Now you just find what you like and chase it with all you have got.

Don't let stress overtake you. If you don't like it, dump everything I said and just do what makes you happy. Don't feel disappointed or rejected that your life got wasted by studying something you are not enjoying. The engineering ship maybe sinking but you can still swim.

What to do if you have Graduated from an Engineering College but want to Shift?

Ignore the cynics and critics: No matter who says what, it is not necessary for you to have a career in engineering just because you have spent lakhs of rupees and wasted four years of your life, to get those two letters 'B.E.' by the side of your name.

Some of the questions that will be thrown routinely at you are: 'why did you study engineering if you wanted to be a photographer?', 'why did you waste your parents' money if you wanted to be a writer?', 'you could have told your parents at school itself that you wanted to be a fashion designer', and the best of all is 'why did you study well at school if you didn't want to be an engineer?'.

These questions are not asked with your happiness in mind. For parents, they won't want the investment in your engineering degree to be wasted and they would find it embarrassing to tell others that their son/

daughter wants to be an artist or a fashion designer after studying engineering. If we consider society which includes your teachers, neighbours, friends, colleagues and relatives, their response is based on what they consider to be 'normal'.

Whether you will be happy with your life 20 years down the lane does not matter to most of these people, including parents in many cases. This is where we need to remember that these cynics and critics are part of the same social system that undermines a child's interest and potential and shapes the child's brain to become a robotic follower of what everyone does.

If you had told these same people that you wanted to be a photographer while you were in your school, they would have laughed at you and said 'Engineering is the future. If you study engineering, you will have lot of scope.'

***

Look at your classmates: If you look at any engineering college's placement records, including IITs and NITs, you will notice a huge chunk of students from various departments being placed in I.T. companies. Everyone from chemical engineering to mechanical engineering compete with computer science engineering students for software coding jobs. No matter how much energy the electrical engineering students wasted in designing circuits or mechanical engineering students in learning automobiles, many end up standing before the I.T. interview queue and learn coding and de-bugging, which is entirely unrelated to what they studied, unless they are in computer science or I.T. I have waited in that queue. For three years, I racked my brain trying to understand Kirchhoff's laws and capacitance values, only to start studying oracle and java at the end of the third year when the campus placements began.

Even the computer science and I.T. students are not an exception. They also prepare for CAT, XAT, MAT and BAT and pursue MBA in finance and marketing, which is again completely unrelated to their under-graduate college study. When a chemical engineering student chooses a coding job at an I.T. company, it is considered normal but when an electronic engineer takes up cinematography it is considered abnormal and strange, even though the career deviation is the same in both cases.

***

Now that you have finished your engineering degree, there is no point in regretting it over and over again. Look at the field of arts. Here lies the best example that your degree need not be a deciding criteria for following your interest. With interest and talent, any engineer can succeed and triumph in a non-engineering field.

Many young actors born to popular actors of the 1980s and 1990s period have studied engineering but later took up to acting. They emerged as talented actors, singers and directors; some have even won awards for their acting prowess. These actors already have lot of money. Unlike most others they had the socio-economic support system to allow them to safely experiment with cinema. If their interest is in cinematic arts and acting, they could have studied an under-graduate degree in any arts field or area related to cinema but like the rest of the crowd, they too joined the engineering train.

***

The best thing about studying in 21st century India is not the number of engineers and doctors who are graduating every year. It is the availability of opportunity to convert hobbies and extra-curricular activities into a full-time career. Many engineers in their resume enter hobbies to impress interviewers when they actually want a career in that hobby.

We write 'Photography' as a hobby in a MBA application form, when we want to pursue photography, not business management as a full-time career. We say 'Singing' is an extra-curricular activity before an I.T. interview panel, when in reality, we want to take up singing as a full-time career. If you give them one crore rupees and tell them to do what they like, they would happily throw away the job and turn to their hobby full-time.

At this stage in life with an engineering degree in hand, either you would have finished your college degree with a non-engineering plan in your mind or you would not have yet decided what to do in life. The common escape route chosen by engineering graduates is either going for a post-graduate degree or starting a non-engineering venture, the current popular word for it being social or business entrepreneurship.

***

POST-GRADUATE DECISIONS

A post-graduate degree is where you can correct the mistakes you have made in your under-graduate degree. Even after recognizing their error and realizing that they studied engineering probably only to satisfy their parents, many repeat the same mistake while choosing their post-graduate degree and post-graduate college. They follow the herd in either preparing for Masters in Engineering or Masters in Business Administration.

When someone is struggling without any interest to study even an under-graduate course in engineering, how can that student be expected to be happy in studying a Masters course in engineering?

When you know that you lost your individual decision making ability by following others in taking an engineering course, why do you still commit the same mistake in following others to a CAT preparation classroom or a MBA school? Poaching institutes popularly called coaching institutes that publish their CAT and GATE discount classes in newspapers seem to be indirectly saying 'now that you have wasted 4 years in B.E., waste two more years in M.E. and M.B.A.'

I am not going to argue against opting for a Masters degree in engineering or a Masters degree in Business Administration but I am warning you not to decide on a post-graduate degree when you are not sure about what you want to study and especially when you are yet to take a decision on what you want to do in life. Some advise that 'first you join a post-graduate degree and everything will fall in place'. What will happen if you realize after joining a PG course in Systems Engineering that you want to do a course in Visual Arts? How will things fall in place then?

The first thing you have to figure out is what interests you in life, irrespective of whether it is arts, cinema, fashion, business administration or technology. Then try to understand whether you wish to work for someone with a post-graduate degree or you want to start your own social or business venture with a post-graduate degree. If you are ready to spend two years of your adult life for a Master's degree in college, then it should not be problem for you to spend two weeks to find out why you like that particular course and what you wish to do in life with it.

The more you learn about your choice of career, more doubts you will have and when you try to clear those doubts by seeking better answers, you will get a better picture of why you like that particular field.

Personal Note: Stop justifying your dreams to your friends and relatives. It doesn't matter whether their cousin is in I.I.M. or uncle is in I.I.T. They should only care whether you are happy and not whether you applied for a Masters in Embedded Systems or Movie Editing.

***

There is another hypocrisy in our system. An arts graduate cannot shift into bio-medical engineering or electronic engineering but any type of engineering graduate can apply for most of the Arts college courses, from fashion to pottery. Isn't that strange? Anyway, this is an advantage and an opportunity for engineering students who are looking to change their career paths. The best thing about the education system of post-2010 India is that there are now many courses available from animation to healthcare administration to tourism studies and the best thing about today's entrepreneurial culture is that you don't require any specialized post-graduate degree for most business/social ventures.

In this scenario, it is unfortunate that some do a MBA degree after their B.E. only to justify starting their own business venture later. They feel that if they start a business with just a B.E. degree they won't be able to justify their deviation and if they add a M.B.A. to B.E. and start a business they can justify it easily if anyone asks.

For a vast majority of start-ups and new business ventures, an MBA is not required and that is usually the first bad investment they make in their business. Instead of doing a MBA degree to open a photography school, you can do a Masters degree in Arts or any photography-related field. This way your knowledge and talent will improve. If becoming a photographer is your sole interest in life, what will you achieve by joining an ordinary MBA college which is not even in the top 20 and even if you join an institution in the Top 20 the college fees will cost around 10 lakhs minimum, with which you can set up a photo-studio and purchase the best camera equipment.

An investment in a website will cost 10,000 rupees and Facebook ad promotion will take another 10,000. With a small portion of your MBA fees, you can lay the foundation for a small company. Sometimes you may have to work if you wish to finance your post-graduate college expense from your own pocket or if the course you are interested in mandates work experience. For many non-engineering Arts courses the fee is not as high as it is for MBA and M.E. Get the fee structure of the college where you want to join and figure out whether you can afford the fees on your own or you will need to work for a while to pay it. Plan it beforehand and you have a better chance of admission and integration with the course if your area of work is related to your post-graduate study.

***

100 % campus placements happen in less than 5 % of all the engineering colleges that exist in India. Often tens of thousands of engineering students from middle-class and poor family backgrounds are unable to get placed in any company during campus interview even though they want and need a job. These are students whose parents would have invested a significant portion of their life savings in their engineering college fees and when the students fail to get a job after finishing engineering degree, it pushes them to depression as well as desperation. When a student with a job offer or a work experience takes up M.E. or M.B.A, he/she has the power of choice but for students who missed their campus placement, their decisions are more likely to be influenced by what is seen as the 'normal path' after

engineering. They are more likely to register for a M.E. or M.B.A. course without any interest and with no sense of purpose. They see it as an escape route to avoid the phase of unemployment after college.

If you are unemployed because you could not get placed during campus interview, don't immediately jump towards a post-graduation degree. It is understandable if you find it very embarrassing and humiliating to sit at home when your neighbour wakes up at 5 in the morning every day and repeats a daily work routine, albeit without interest. This unemployed phase is where many make mistakes. To escape from the problem, they immediately register for a M.E. or M.B.A course.

At this stage, it is necessary to figure out why you are unemployed in the first place. Is it because you don't want to work as an engineer and really want to do something else? Then it is very good. Don't bother about what your neighbours think and just start focusing on your plans. If it is because you want to work as an engineer or manager but could not get placed in college, it is better to spend some time figuring out whether you missed placements because you did not have the skill or your marks were poor. I am not trying to discourage you. The hidden fact is that the campus placement market remains the same at B.E. and M.E./M.B.A level in most of the non- IIM and non-IIT colleges. The same companies come for campus placement for B.E. and M.E./M.B.A graduates. If you missed placements because of poor English or communication skills, which is the common reason for many rejections, then it is better to focus on that before joining a M.E. or M.B.A.

In the mean time, shed your ego and try to work at any job that is available after college as a stop-gap arrangement, till you find clarity on what you want to study. You can do a M.E. or M.B.A even at the age of 40. It is better to study with clarity at the age of 28 than with confusions at the age of 21.

ENTREPRENEURIAL PATH

Two things about life which every change-maker should realize early in life are: New ideas soon become old and there is no such thing as the 'right time'. Right time is when you are ready to implement your idea and we are never fully ready. Sometimes you just have to start after doing the basic preparation and become fully ready on the way.

Any business or social idea has three dimensions – Plan, Finance and Marketing. How you combine them is your mode of operation. That sounds like an MBA-type speech but the good thing is that you don't need an MBA degree to figure that one out.

Two common worries for those starting their business or social venture are i) whether a post-graduate degree is required to understand the business and social process and ii) whether work experience is required to gain knowledge before starting your own company.

Decide whether the work you wish to do to gain 'experience' and the post-graduate degree you wish to do to gain 'knowledge' is relevant to your business or social venture in the first place. In many cases, you do not require a post-graduate degree or work experience for launching your own start-up and there are lot of entrepreneurs who waste their time with a wrong combination of work experience, post-graduation-degree and start-up idea.

Don't mindlessly repeat clichés like you wish to work for few years for 'experience'. Tapping away at a computer in a cabin is not the relevant experience for someone aspiring to be a photographer, writer or social activist. If at all you require experience, try to find it in a company that is relevant to your dreams and long-term plans. If your work is relevant to your future plan, then you can fruitfully use your work to gain real experience.

***

You do not require a general MBA degree to start your own app-coding company. You will benefit more from a technical degree in computer science or information technology. Similarly, you do not require experience in an I.T. company when your ultimate aim is to take up photography full-time.

If your idea is to take up writing full-time, then try to work for a newspaper. If your idea is to launch a fashion store, then try to work for a fashion-related group. For many who are starting a non-computer related business or social venture, working in an I.T. company seems to be the default and sole option for gaining 'experience'. This kind of experience is required to fill the 'years of experience' column while applying for M.B.A. or for work but not for starting your own business.

On the other hand, it is also painfully true that most of the employment opportunities for engineering graduates is in I.T. So if no other option is available, just use the I.T. career to meet like-minded people, travel and earn money, and try to see whether any knowledge related to your own future plan can be gained from it.

***

If you are working or wish to work only for financial reasons to finance your idea, then you will have to put in place a timeline containing the total months you have to work to earn the investment required. If your monthly salary is 25,000 and your monthly expenditure is 10,000 you will require at least 10 months of work experience for a 1.5 lakh investment. If the money required is less, this could be a better option than standing in front of an Angel investor or a loan manager.

Money does not play a big role on its own. But money in the form of finance plays a crucial role. You need money to finance your investment, travel, advertisements, equipment, plans and Facebook ads. If you are planning a new venture or business, see it in proportion to your college fees. If you had spent 3 to 5 lakhs in engineering college admission fees and given 5 to 10 lakh in management fees, spending few lakhs on your idea is not really a big investment to worry about.

For personal reasons or family situation, some may not be able to quit their jobs and focus full-time on their start-up business or a social cause. You don't have to quit your job but don't quit your dreams either. This will require lots of sacrifices but if you believe in your dream strongly, you can make it work. While you are working, start focusing on your business idea on a part-time basis till it shows signs of potential profit. If you have a trustworthy friend, make him a partner so you both can share the burden.

If it is a business idea which you soon intend to convert into a company, then you can consider moving to a low-paying but less-stressful job which gives you time to focus side-by-side on developing your idea and turn it into an enterprise.

If it is social venture plan that is not motivated towards earning money and if you only wish to use your idea to help the society, you can also consider directly working for an established social service group or an NGO which is involved in charity spending. This way you earn money and also serve the society. Many IT and engineering companies use their Corporate Social Responsibility programs to involve their employees in social service. Some companies involve their employees in building schools, while some are involved in health campaign. You can check their CSR programs and join the company whose program matches your social idea. You will gain field experience about how social change works on ground and also the money to later start on your own.

***

Nishanth wanted to start an educational NGO for slum children. As a beginner, he could not find any major investors who were ready to back him for few years. He decided not to quit his job and managed to convince the manager in-charge of his company's Corporate Social Responsibility program to run a week-end coaching class for the slum children. This way he was able to both stick to his job and work for children.

***

If you are only an under-graduate degree holder, then start your business or social venture after college and if it fails, you can always do a post-graduate degree and get back into the job market. Irrespective of your educational qualifications and work experience, ensure that you register your business or social venture and file tax returns, so you can claim it as work experience for both higher study and future employment.

***

It is your wish if you want to do a Strength-Weakness-Opportunity-Threat analysis about your ideas and dreams but do understand where you lag behind. Lagging behind is not a negative point. No one is perfect. You can always hire someone to do what you cannot do. It is enough if we have an idea of what has to be done.

You don't have to know how to design your shop's website. It is enough if you know what it should look like. Then hire a designer and get it done. If you have an idea for an app and you don't know coding, just hire a software engineer to do the coding with your ideas.

A fashion designer friend of mine had lot of saree design concepts but she could not even draw a circle. So should she have given up? No. She did the obvious. She hired a team of artists to do what she couldn't do. If you look at the story behind many successful websites, including Facebook, the leaders always hired people when they know they were poor at something. They understood what their short comings were over a period of time and they simply asked someone else to fix it.

***

Planned risk, not blind risk: The 'risk' that you take in a physical adventure activity like trekking and exploring is different from the 'risk' that you take in career planning. It would be a blind risk, if you take a decision without knowing why you are taking it. Plan your risk after satisfying yourself that this is what you want to do. You may fail. You may succeed. That is not important. What is important is knowing that this is what you want to do.

Even if there is a fear of uncertainty over what happens in case you fail, understand that sometimes taking a risk in pursuit of our dreams and suffering is better than regretting not taking the chance all your life. It is better to fail at doing what we like than win at doing what we don't like.

***

Like it or not, we live in a society which will look down at you if you start a chocolate firm, after finishing engineering; it is seen as degrading to your engineering background. But the same society will praise you if you finish a business degree at IIM and work for a leading global chocolate company. This is how social hypocrisy works. So don't care if your start-up idea or career plan is not related to what you studied in college. And don't be worried about what everyone thinks about it. Just get on with it. The applause will come when you succeed, but to succeed you need to start somewhere.

For those still in School

Let me begin by congratulating you for being in the best phase of your life. This is a phase of life which you can enjoy only once. Though what is theoretically seen as school life extends from L.K.G. (lower kinder garten) to 12th standard, practically for most school students, school life ends at the end of 8th standard itself. While schools don't have corporate campus placements to boast of, they derive their pride from the number of students who top the 10th and 12th standard board exams. As the number of students who score above 90% increases, so does the brand value of the school and the school fees. The quality of a school is not linked to the quality of its teachers. It is linked to the quantity of marks their students get.

You would have seen how some students who score consistently well in school's internal exams are treated 'special'. In certain ways, this is how life will be too. (I have been such 'special' student once as well.) This special treatment comes from what is seen as 'success'. Here success in life and in school is equated with marks. But marks don't help you in the long run, if you are not happy with your life. I am not going to say that you should drop your books, bunk your classes and score as low as possible, (though it's not a bad idea to bunk classes sometime and have fun). As far as school exams are concerned, do your best without stressing your mind and body too much. Try to score as many marks as you can.

There are two important career decisions a school student has to take which may decide the career progression of his/her entire life.

1) Choosing commerce or science stream after 10th standard.

2) Choosing under-graduate course after 12th standard.

Commerce or Science?

Most state boards, central boards and ICSE syllabi include English, the state language, Maths, Physics, Chemistry, Biology, History, Geography and Ethics, up to the 10th standard. After the 10th, the subjects are grouped into various streams. English and the state language are common for all streams. Almost all State Boards and CBSEs streaming of subjects fall in the following categories:

Science stream 1: Biology, Physics, Chemistry, Mathematics.

Science stream 2: Computer science, Physics, Chemistry, Mathematics.

Commerce stream: Mathematics, economics, commerce, computer science.

The commonly found scene in most schools is that the students who study well in 10th standard are offered, motivated and many times forced to get into one of the science streams and the students who don't study well are given commerce stream. The students are not consulted while making this choice. Instead the choice is made by teachers and parents. It is understandable if parents, who are not professionally trained in child psychology are unable to identify the unique quality in their children. It is shocking that even teachers who should know better than others that each child is unique, play an active role in destroying the individuality of the children and just sort them into science and commerce streams based on their marks and not based on their interest.

Another limiting factor is that the schools will have only a fixed number of seats for science and commerce studies and they quickly fill the science seats with top students and once the numbers are filled the remaining students have no other option than to join commerce stream if they wish to study in the same school. Sometimes priority is given to other school students who scored higher than their own school students for admission to 11th standards. Ideally the progression from 10th standard to 11th standard should be a continuous process but schools run it as a fresh admission for 11th standard for both their existing and new students. (Students who studied in a school from their 1st standard are made to apply fresh for joining 11th standard in the same school.) This is like a mini-college admission process.

Schools rarely have any career counselors and many parents have no idea about where their child's actual life interest lies. If you are in a position to convince your parents about what you wish to study, then you can base your decision using the following criteria:

College course interest:

There are some who are very sure that they want to be doctors or engineers and others may want to try their hand at business or CA.

Life interest:

You may not know what to study in college but you know that you want to be a photographer or a painter or a movie star, when you grow up.

***

Science (Biology + Maths) is the stream using which you can go for any type of college course, from medicine to commerce to B.A. History. Whereas with commerce stream you will not be able to study medicine/ engineering related college course after 12th standard but you are eligible for most of the arts courses.

If you have the academic interest and capability to study science and maths and if you are not sure about what to study after 12th standard, you can consider the science stream. This way you will have wider options to choose from for your college studies, options that are as diverse as B.E. Textile Engineering to B.Sc. Visual communication.

If you strongly believe that you don't like or cannot study science and want to go for only arts or finance studies, then you can consider the commerce stream.

If you are already very sure about the college degree course you want to take after 12th, verify the school stream that is required for you to be eligible to take it. Some courses require biology and some courses require mathematics. While many colleges are flexible with the marks, unfortunately the stream chosen for 11th and 12th cannot be changed later.

What to study in college?

When we are small kids, if uncles and aunties ask us what we wish to do in life, we give an honest answer that comes from our heart: 'Uncle, I want to be a cricketer.', 'Aunty, I wish to be a painter.', 'I want to come on television every day.', etc. As we grow up, (approximately around the time of 10th std), our answers come not from our heart but from our brain. Our answers sometimes already decided by our parents and relatives. 'Uncle, if I get very good marks, I will go for medicine. If I get good marks, I will go for engineering. If I don't get good marks, my father threatens me that he will join me in commerce.' I have been there. Believe it or not, after encouraging me to pursue engineering all my life, my father actually made me apply for a pharmacy course when he feared that I won't clear the engineering cut-off.

***

Your interest in a college education will be based on your world view and the social conditions around you, which are usually set by parents, school and relatives. Very few school children will be able to take a under-graduate college course that matches their world view and life-interest.

An understanding of both the available college courses and your life interest will help you to use your college life as a launch pad to attain the life you want to lead. So it is important for you to first decide what you want to do in life and then check the available college courses that match your life-interest.

***

It is always good to study what you want to become. Study economics, if you want to be an economist and study arts, if you want to become an artist.

If you want to be a photographer, artist, movie-maker, actor or model, then you can go for courses like Visual Communication, Arts, Mass media studies, etc.

If you want to work as a banker or a stock-broker, you can go for economics or commerce.

If you want to be an I.A.S. officer, social activist or become the future Prime Minister after starting your own Aam Aadmi Party, you can try law.

You will be spending 3 to 4 years of your late teenage and early adult life in an engineering or Arts and Science college. You won't lose much if you spend 3 to 4 weeks before college admissions, trying to figure out what you want to do in life.

***

If your parents are the 'strict' and 'disciplined' types for whom their choice is more important than your interest, then you may have to sacrifice atleast the under-graduate college degree choice.

You can first try to reason with them, argue and even cry that you don't want to be pushed into a college course that you don't like. However, if they still insist, don't be disheartened or sad. Cry for a while if you want to. But once your tears have dried, realize that the same engineering college system that you now hate has also produced more popular artists, photographers and writers than arts colleges. You can still pursue your real interest even after college. You may also find a new passion and a new set of dreams during your college life.

If your parents are liberal and if you have an idea of where your interests lie, then you can consider seeking advice from a career counselor or from people who have pursued courses related to your interest. A career counselor charges around 1000 rupees per client. If you are ready to invest 5 lakh rupees in an engineering education, you can very well afford to pay a very small percentage of that amount to a career counselor and understand your real academic and life needs.

***

Focus on the subject, instead of the degree: For example, if your interest is in chemistry and if you are not able to get a chemical engineering seat in a good engineering college, you can opt for B.Sc. Chemistry in an equally good science college. There are some who join other engineering courses like electrical and mechanical engineering instead if the chemical engineering seats are filled up. If your interest is in chemistry, you can still find employment in the chemical industry by pursuing a Bachelors course in Science. A B.Sc. degree in chemistry from a reputed arts and science college has more value than a chemical engineering degree from an ordinary engineering college.

If you're sure about the field but not about the area of specialization within that field, then go for a general course in your field of interest instead of a specialized course. If you want a career in the field of cinema, but you are unable to decide whether it is photography or direction or editing and if you are confused between courses like B.Sc. film technology and B.Sc. Visual Communication, then go for B.Sc. Visual Communication, which is a general course dealing with all areas of photographic and cinematic arts. During the under- graduate course, you can experiment and try your hand at different areas of cinema and for post-graduation you can choose your specialized area of interest.

***

There are some integrated programs like the 5-year combined degree (3years B.A. + 2 years M.A or 4 years

B.E. + 2 Years M.E.). Avoid such courses if you feel that your life interest will change after few years. And if engineering is not your passion, don't even look at them.

***

There is no scientific classification of physics as IIT physics, engineering physics and B.Sc. physics. Physics is just physics. Educational institutions have cut physics into such pieces based on their view of students' ability to understand physics. It varies from one educational sector to another. There is a myth among the educated population that IIT physics is greater than Engineering physics, which is greater than B.Sc. physics. Among all these categories, it is usually the B.Sc. physics student, who studies physics properly, would have joined the course out of interest and will pursue a career related to it.

There is a good chance that the student who studied IIT physics will end up working as a stock broker and the student who studied engineering physics will be a banker. Most likely, it is the B.Sc. Physics student who would have gone on to pursue a Master's degree in Physics and attempted a career as a physicist.

***

When you are being forced to join engineering against your will and if you have no choice other than to join one, then choose it based on these factors: The companies that come for placements, sports facilities in college, extra- curricular activities related to your artistic interests, freedom on campus, and the quality of college life. There are several engineering colleges that are culturally rich. Their cultural festivals are more famous than their technical symposiums. If you are sure that you want to join a non-engineering field after finishing engineering, then you could join these 'cultural' engineering colleges. Join an engineering college that is famous for its cultural festivals and allows students to attend as many cultural festivals as they like. Frankly speaking, it is better to join the best college available that does not care about what you do.

If your interest is in science and if you have joined the wrong branch of engineering, due to lack of availability of seats or other reasons, you can still pursue your original interest through your Masters. There are masters degrees in engineering from robotics to bio-medical engineering. Use your free time in college to prepare for GATE and GRE. You are still lucky, at least you won't have to face the kind of criticism that engineers turned artists face.

***

The father of a friend of mine owns hectares of tea plantations and rubber trees in the hills of Kerala. They also own a few houses in Bangalore's suburbs. He is financially secure and did not have any interest in engineering or even science. After school, all he had to do was decide what he wants to do in life and do it. He had studied science in school so when he told his father that he wanted to study journalism in college and take up travel-writing, his father just laughed at him and put him in a premier engineering college through management quota, by paying large sums of money for the admission fee. In his family, any major decision is taken only after consulting a specific group of relatives and all his relatives felt that studying in a top engineering college is a must for anyone going to college from their family. Now he is working as a junior engineer at an electrical company (a job he doesn't enjoy) after finishing engineering (a course that crushed his enthusiasm).

For many rich parents, their children joining engineering is not an economic necessity. It has more to do with their social status. While, for most middle-class parents, admitting their children in an engineering college is an opportunity to rise above their middle-class status through the income of their children.

***

Critics would argue that it is shameful to live off our parents' money and that we should give back to our parents the money they spent in raising and educating us. In my eyes, however, this view makes the entire parent-child relationship look like a business transaction. Perhaps that is how they see careers and education too. The best thing that any loving parent would want is for their child to be happy and the best way to make any loving parents happy is by doing something we enjoy and excelling at it.

***

On one hand, those who want to become engineers cannot afford to pay the management fees and on the other hand, those who have enough money to even buy an engineering college, join engineering without any interest.

There is a clear separation between rich and middle-class because the need to be economically successful is high in middle-class families and it adds additional pressure on children.

One of the many things children from middle-class families would have realized by now about life is that it is economically unfair. A lot of money is in the hands of those who don't know how to spend it and only a little money is available for those who can make better use of it. It is common for middle-class parents to impose their ambitions and dreams on their children by making them engineers and doctors. These parents would have stood in queues for kindergarten admissions, raced to secure high school admission forms before time and would have slogged for several long hours at their workplace just to get a bonus that will bring extra rupee to pay for your engineering college fee.

While all this is very touching, the ultimate question is whether you should pay them back by sacrificing your dreams, by studying a course you don't want to study, by pursuing a career you don't like and by living a life you hate? The answer, I think, is resounding NO. Unfortunately, they may not accept 'no' for an answer after wasting 20 years of their lives just to see you become an engineer.

***

If your parents are strict and are determined to make you study what they want you to study, then you may have lost the battle but not the war. Don't lose hope just yet. There are wildlife photographers with an automobile engineering degree and bikini models with a chemical engineering degree. You can still reclaim your dream by doing a post-graduation in the field of your choice.

There are also many others who use their engineering degree as a launching pad for their non-engineering plans. But let this involvement of parents in your life be the last. Taking care of parents is different from being their slave. Ensure that after college, you are able to decide your life choices based on your life needs alone.

***

Study what you love, not what gives you money.
Study a degree that increases your happiness,
not the one that increases your bank balance.

Four Year Course for a Five Digit Salary

If there are no campus placements and the guarantee of a job at the end of a four-year engineering course, you wouldn't find many takers for the engineering degree. Even at I.I.T., if the government makes an announcement that job placements cannot be guaranteed after graduation, parents will lose enthusiasm in making their 14 year old children wake up at 4 in the morning to rush for IIT-JEE coaching classes.

An engineering degree is seen as a financial and social investment. Parents deposit their children with few lakhs of rupees in an engineering college and at the end of the four years; they expect a simple interest in the form of a good campus placement. This is where I blame the Arts and Science colleges. They have failed to be as aggressive as engineering colleges in bringing companies to recruit their students.

Now when the engineering student gets placed in a company, the money invested in his education can be recovered in just few years. This is much quicker than it is for Arts and Science courses, where graduates usually go for a Masters degree before looking for work related to their study.

When I got placed during my third year of college at one of India's premier software companies, the training salary offered to me was around 21,000 per month. (Those glamorous 1 lakh per month salaries are mostly for IITs) Still 21,000 did look like a big figure because I hadn't yet earned my own money or enjoyed the pleasure of spending what you earned. But it did make my parents happy and relieved that their investment had started returning profit. Those parents whose children couldn't get placed would have ended up feeling that they could have saved a lot of money if their child has joined an Arts college instead.

Ever since my campus placement, whenever I meet someone who holds a post other than as an engineer, be it a watchman duty or an air hostess, I have developed a habit of asking them their salary. (From my experience, I suggest you ask when they are alone.) Their answers have allowed me to compare the engineering job market with other individuals around us.

***

A night-shift security guard at an I.T. company where my friend worked for a short-time earned 7,500 per month. If you are an ex-military serviceman, you can get a maximum of 5,000 rupees more, he said. My friend's salary itself was Rs.15,000. This guy had an engineering degree but the watchman who just passed out of 10th was earning as much as him.

***

At a famous juice shop, where I drink after my daily run, there is an old man who stops the traffic on the road temporarily, so visitors to the shop can reverse their cars from the parking lot and move easily to the road. His work, which may or may not be a part of his official duty as the gate keeper, gets him a tip of 5 to 10 rupees. About 50 to 100 cars come in a day and he gets at least 250 to 500 rupees a day, which touches the monthly salary, my friend was getting then.

***

While writing this book, I was working as a financial assistant at a national auditing organization. My job was to look into tax assessment details and correspondences related to it, for which I was paid Rs.25,000 per month. While working here, I met a Data Entry Operator, who had cleared the Staff Selection Commission's recruitment exam to get into government service. His monthly government salary was Rs.21,000. He had graduated in English Literature, through the open university, IGNOU. He had not attended the regular day course like most of us but his monthly starting salary is much higher than what many engineering graduates get after four years of vigorous classroom study.

***

Assuming that it costs a minimum of 50,000 rupees per year, the total cost at the end of four years would be two lakhs, estimated at the minimum possible value. A similar investment in a high-definition camera and lens will cost between Rs. 1 lakh and Rs. 1.5 lakh. As a photographer, at weddings or events, you can get Rs.2000 to 5000 per day. Though it is easier said than done, if one wishes to be a photographer or a writer or an artist, the same effort and energy utilized in coding and circuiting can be utilized in their field of interest to do comparatively better.

***

Some spend up to 20 lakhs on capitation/management fees to get a seat in an engineering college, through the management quota. With the annual fees and other expenditure on notebooks and travel, an additional 3 to 5 lakhs can be added. If the same 25 lakhs is invested in Fixed Deposit and Mutual Funds scheme of a nationalized bank, the amount you earn through this financial investment will exceed the annual salary you will get in many I.T. companies.

***

Specialist vs Generalist:

Engineers venturing into non-engineering field do not happen in the private sector alone. Unknown to many, it is happening more in government recruitment than for I.T. companies. In government service, two classifications of posts are maintained — Specialists and Generalists.

The 'specialist' posts are technical in nature and require a specialized degree. For Indian Engineering Service and Junior Engineer posts, an engineering degree or a science degree is required. You cannot work as an electrical engineer with a chemistry or history degree. Whereas for 'generalist' posts like Indian Administrative Service, bank probationary officers, income tax and excise inspectors, any degree-holder is eligible to apply. Here, you can even work as a bank clerk with a mechanical engineering degree.

Many know about CAT exams where engineers deviate to MBA. The new hot favorites for recruitment in the job market are Staff Selection Commission's Combined Graduate Level exam and IBPS' banking probationary officer exam, which are now highly popular among engineering graduates and MBA degree holders. These exams for generalist posts are seeing lakhs of applications from engineers every year. You will be surprised to know how many B.E. + M.B.A. degree holders apply for bank clerk exams. You go to any bank in India. You can easily find an engineer working as a clerk or a probationary officer. A post of a bank clerk is comparatively low when compared to the degree held by an engineer but this just shows the extremity our youngsters are being pushed to when they realize that even with an engineering degree it is not easy to find a job.

***

Many engineering graduates work for less than 10,000 rupees in clerical posts with exaggerated names like Data Processors, Office Support, etc. I was selected for a MNC, where my starting pay was 21,000 and I studied in the second best engineering college in my state! It is apparent that many of those who wasted four years in mechanical, electrical, chemical and all sorts of engineering end up doing clerical work through computers. The work they do and the salary they earn does not match their degree in any way.

If you check the 'Job Vacancies' in the 'classifieds' of Sunday newspapers, you will find many B.P.O's saying only B.Sc. and B.A. graduates should apply but not engineering graduates. Even under-graduates from other courses can apply for part time jobs in these companies but B.E graduates are not allowed. During the colonial rule in India, 'Indians not allowed' boards were kept outside some European clubs. It is a matter of time before 'Engineers not allowed' board is kept outside companies.

***

In the 1980s movie 'Varumaiyin niram sivappu – The Colour of Poverty is Red', actor Kamal Hassan works as a barber in a hair salon, unable to get work as an arts graduate. The movie was remade in Hindi as Zara Si Zindagai. I suggest you give it a watch and you will be able to see how similar being unemployed in the India of 1980s and the India of today is. You will also realize that making money is not the only way of life.

A similar movie made today will probably feature an engineering graduate unable to get a decent job with an engineering degree and struggling to be employed to escape social shame. Now it would be titled 'Face of Poverty is an Engineering Degree'. In the 1980s movie, Kamal worked as a barber after finishing M.A. Philosophy. Today several engineering graduates work as clerks and office assistants with M.E. and M. Tech certificates. In the movie, there is a scene in which a customer at the hair salon runs away from Kamal Haasan after finding out that he is an Arts graduate. It is strikingly similar to how engineering graduates are turned down by B.P.O's saying that they are over-qualified.

***

In a state with over 300 engineering colleges, only the top 10 are considered the best and those ranked in the next 10 are considered 'alright'. It is in these top 10 engineering colleges, we see all the glamorous campus placements, with lakhs of rupees as monthly salary for someone who is yet to even finish their 4th year of engineering. But what do those who study in engineering colleges that are not in the top 10 or 20 do? It is these students who are completely let down by the education system. There are thousands of engineering colleges in which there is no such thing as campus placements and after graduation, the original motivational slogan 'proud to be an engineer' turns into 'let's get any job that gives money'.

Structured Syllabus for a Structured Life

In a four year engineering course, each of the 8 semesters have around 5 theoretical papers and 3 practical papers. The first year usually has the same subjects that you would have studied in 11th and 12th standard, just at a slightly higher level. The subjects are basically physics, chemistry, mathematics, english, etc., with a shiny coating — engineering physics, engineering chemistry, engineering mathematics, technical english, etc.

The 'real' engineering subjects start from the second year, where the in-depth study of concepts unknown to most students earlier comes into play. This is when many start to realize that passing each semester may not be as easy as it was in the first year. Studying the same school subjects in the first year helps the students settle down in the academic structure of an engineering college and helps them slowly come face-to-face with the real engineering subjects, which are obviously not as easy as the basic subjects studied in the first year. Many would have easily scored above 80% in the 1st year but from second year onwards they will struggle to get above 70% and soon just passing would be more than enough for most students, to prevent them from getting disqualified for appearing in the campus interview.

***

Students who consistently manage to score above 85 to 90 % are the ones who are able to study difficult subjects and understand complex concepts; and even if understanding proves to be difficult, they can memorize them easily. Our educational system considers these students as Prize Trophies, as they represent the ultimate role model of how every student should be. Even in the absence of a proper teaching environment, these students manage to clear their papers with flying colours. The usual difference between those who perform well in the exams and those who don't is consistency and regularity. Those who score above 80%, regularly score above 80% and those who score below 70%, regularly score below 60-70%. When the overall percentage drops for the average student while moving from first year to second year, some may think that one or two subjects may have been tough but that is not the case. Low-scoring students consistently start scoring around 70% in all engineering subjects from second year, except for subjects like principles of management, environmental science and engineering ethics, which are really not technical subjects.

After the semester exam results are announced, usually the college principal holds a discussion with the head of department, class teacher and other subject teachers to find out why students have scored low and what can be done to push more students into the 'above 80%' category. These discussions are important to the college as a higher percentage is needed to maintain the impression of 'quality education'. But rarely, is the subject matter being taught, how it is being taught, the quality of the syllabus and the interest of the students to learn the subjects given as much importance as the numerical value of the pass-percentage.

***

Does our syllabus really teach us what engineering is about? Do our exam patterns offer enough flexibility for creative input? Can students decide what electives they would like to study? Only for the first question, the answer is partly Yes and partly No. For the rest, the answer is a straight No.

When a student studies about electrical power generation, the syllabus explains it in terms of motors, transformers, generators, transmission and distribution. Real life case studies of how the electric distribution network works in India from generation to distribution is often left out of the syllabus. Those who study electrical engineering will know how a hydro-electric dam works but they rarely know where the tallest and longest dams in India are located or their socio-economic significance. Students study computer science and Information Technology without even knowing general information like the ministry responsible for taking decisions on software regulation and graduate without knowing the difference between NASSCOM and FICCI. Having a real-life connection allows students to relate to the subjects, even if they are dry and boring. The management of engineering colleges, however, don't realize that.

Forget ministries and agencies, how many E.E.E. and E.C.E. students know the difference between the words 'electric' and 'electronic'? Or, you could ask a chemical engineering student how the chemistry that he/she studies as a part of Bachelor of Engineering is different from the chemistry that a B.Sc. Chemistry student studies as a part of Bachelor of Science. You will be surprised by how so many students graduate without knowing the difference between the two. Let's talk about something that is common for all engineering students and graduates: the word 'engineering.' How many know its origin and meaning? You stand before an engineering college or an engineering college admission counseling session and ask the students coming out the meaning of the word 'engineering'. Very few students would know it. How many parents themselves know what the word 'engineering' means? Many just associate the word with a form of education that provides campus placement and a five-digit salary at the end of it. Our educational system has made us blind to the obvious facts of real life and instead focuses on knowledge that lies within the pages of a five-hundred page textbook.

I, for one, found out what the glorious word 'engineering' meant only when I finished my 2nd year. I then realized that I'm really not an engineer at all.

***

A computer science student who learns about the origin of Infosys in classroom will be inspired to start their own firm after graduation. A bio-technology student who learns about how Biocon was founded and how it found global dominance will be motivated to start their own bio-tech venture. When I was studying, Yahoo Messenger and Orkut were the popular internet tools (they were the Facebook and twitter for those who studied before 2010) but their coding or design systems were not part of our syllabus. If it was a part of what we studied, then the course might have been more interesting and relatable . Even now, Facebook is yet to be made a part of computer engineering education. If Facebook's coding platform, its origins and its rise to the top can be taught to our students, it will help them combine both theoretical and practical knowledge in the most real-life scenario possible. This will also motivate them to dream big and create such companies rather than just being satisfied with a well-paying job. Motivating an engineering student to start the next Facebook is more important than getting them to work for Facebook.

***

'Practicals' are a great way to learn, understand and study what is taught in theory. But very often both the practical and theory paper on the same subject are taught like two different subjects, by two different teachers from two different departments. Both the teachers are concerned only with ensuring a high pass percentage in their subject and very few teachers take the effort to connect the practical and theory subjects. The syllabus pattern does not allow it and hence it is up to individual teachers to decide whether to link both.

Electronic Circuits is a theory paper in third semester and a practical paper in fourth semester. For some theory papers which require hands-on experience, there won't be any practical papers. For practical papers that require bookish knowledge, there won't be any theory paper. If in a semester, students could study 3 subjects in sets of pairs, like Electronic Circuits (Theory, Practical), Java (Theory, Practical) and Mechanics (Theory, Practical), keeping the focus on making students touch, feel and explore what they study in books, the syllabus can be made a lot more interesting. Even in the examination pattern, a combination of theory and practical exams for each paper can be considered. We have so far maintained a separation between classrooms and laboratories. Labs are the temples of learning and ideally labs should be classrooms. This also removes the boredom that comes from sitting in the same place all day long.

***

In the answer booklet of exams in some universities, there are fixed spaces for answers. For two marks, the space has only five lines and for 16 mark questions, the space goes for two to three pages. Your answer cannot exceed that space limit pre-fixed in the answer booklet. It makes one wonder whether the examination is meant for students in kindergarten, where L.K.G. children are taught to write the alphabet within the box or space provided in the test sheet. Engineering students should be encouraged to write their own answers, based on their own understanding of the subject instead of blindly memorizing the answers given in books. Different students have different answers to the same question. If a student has a better answer than the one provided in the standard book, there is no way that student can showcase their independence of thought in the examination paper. The student will have to dump their individual theory for the method given in the book, because only that will provide a solution within the fixed limit of two pages and fetch them the marks they need to pass the exam. Such limitations imposed in the answer sheet limit their creativity and also prove the theory that our education system wants us to think only within the box, or in this case, within two sheets of paper.

***

Engineering education is tailored to suit the demands of the private industry, to satisfy the demands of your future workplace. Industry-Academia meeting are conducted, informally and formally, between education officials and industry experts to discuss how the education should be updated and modified to suit the current and projected industry requirements. This is entirely needed and appropriate as majority of the students aspire to get placed in the industry and industry leaders should be kept informed of the on-going changes within the educational system. These Industry-Academia meet will be greatly enriched if the third stake holders of the process— the students are allowed to voice their opinion as well. No matter how mature students think they are, many may not have the long-term vision to see how engineering education plays a role in industry growth and development. But industry leaders can get a better insight into the educational system when they directly interact with both the exam-topper and low-scorers, without the looking glass provided by the universities and education departments.

***

India produces a lot of engineering graduates but not engineers. An electrician knows more about electrical wiring and installation than an average electrical engineering graduate. A mechanic knows more about bikes and cars than an average mechanical or automobile engineering graduate. Even if a student has taken up electrical or mechanical engineering without any interest, giving the student hands-on experience in electrical wiring and operations of car mechanics will create interest in students to explore his/her area of study.

When I go to my village for summer vacation, the people there look at me as a proper engineer, rather than an engineering graduate. Once my uncle asked me to fix the wiring for a newly installed electrical irrigation pump. For them it was a matter of pride to see an engineer from their own family to do the work for which I studied. I had to admit with shame that I only learned electrical wiring in books and that I had never installed any wiring on my own. Then we had to call an electrician who then set it up. The electrician looked at me with respect when my uncle told him proudly that I am an electrical engineer. Fortunately, the electrician didn't ask me why I didn't repair the motor pump. Hiding my embarrassment, I openly told that he knows more about electrical engineering than half of my electrical engineering class. What I studied for four years was of no use to me on the ground. What I studied in books, stayed in the books.

***

Many engineering students will be surprised to know that the most commonly used computer application across all industries from User Operations to Information Technology to Merchandising is not C or C++; it is Microsoft Excel. Its importance is grossly under-estimated. We studied it as a four page unit in a chapter called MS applications. We did not even study it as a separate subject and yet almost everywhere MS Excel is the cornerstone of computer operations.

***

Our syllabus is not designed or implemented to encourage original thinking; it encourages memorizing facts and theories. Any creative and free-thinking student who likes engineering would find it difficult to stick to this pattern of education. As a result, the course of study not only fails to generate interest in students who have chosen engineering under force, but it also actually destroys the enthusiasm of students who joined engineering with interest.

***

How many marks you get matters more than how much you learned. How many papers you cleared matters more than how many concepts you understood. Most of all, how much you earn after college comes to matter more than how much you learned in college.

Changes Needed in the System

1) Counseling: This is the most hypocritical and idiotic word used in our education system. The word counseling means 'professional assistance and guidance in understanding personal problems and trying to arrive at a solution'. In the context of education, it means 'helping the student understand what he/she wants to do in life and find a suitable college degree course that best matches the student's dreams and desires'. This is the meaning of the word 'counseling' in countries, where their engineering and business schools top the world ranking.

   Now, what do we do in our 'counseling' session? We just compare our marks in the rank list and choose the most suitable engineering and medical college available on the day of counseling. We basically sell our students to colleges based on their marks. When I went for my counseling session, there was a time-limit within which I had to choose my college or some other student might choose it before me. As usual, my father made the choice for me and my 'counseling' was over, in less than 60 seconds. Those who run the engineering counseling process are not only ignorant about engineering education but also about teenage psychology. Teenagers think a thousand times before choosing the clothes they wear. Here, they are made to take a decision that will affect the rest of their lives without even thinking twice.

***

Children should be educated, at a young age, that making money does not equate to being happy. Career counselors and child psychologists should be appointed at school level to explain to the children the various courses and options they can pursue that match their life interests and skills. So much money, resource and energy is spent in creating national employability programs for unemployed engineering graduates. A fraction of it if spent on a national career plan for school children, to help them find what they really want to do, would lower the unemployment levels greatly, while also making sure they are happy with their work. Many engineering graduates are unemployed because they joined the course without any interest and now they don't know what to do with their degree.

If those who don't like engineering stop taking engineering, then the seats at prestigious institutions can be filled by students who have an actual interest in studying engineering. We don't see much progress in inventions in India, because these students are studying what they don't want to study. When interest meets purpose, we can see some progress in the quality of our engineering talent and see new discoveries made by our engineers.

If a child has interest in medical science but couldn't secure enough marks to get admission to a M.B.B.S. degree, then instead of making him join a random engineering course, the child should be made aware of similar courses like B.Sc. veterinary science, Dentistry, B.Sc. zoology, etc; and if his interest matches with any of these alternative courses, the child can join one of them. Even if engineering is the only alternative, then courses like bio-technology or bio-medical engineering can be looked at if they still wish to pursue their interest in medical science, without an M.B.B.S. degree. Let the student choose among related options based on their interests. One of my many dreams during school was to work for the National Geographic magazine. I love animals a lot and I wanted to travel to different natural wonders of the world. If I knew a course like B.Sc. veterinary science existed when I finished high school, I might have chosen it. Instead, I was given only two options by my parents — engineering or medicine and years after graduating, I am yet to figure out which option would have been worse for me.

There is some form of 'counseling' that exists in many schools but it is a pure mockery of the psychological concept of counseling. Instead of focusing on why the student's marks are lowered, we should find out whether this person is happy with their life or not. That alone will solve most of the problems of childhood and adolescence.

2) Every college student has a set of dreams and a set of confusions. Career counselors should also be appointed in colleges and universities to guide them in the direction they wish to go, instead of letting them get pulled into the routine motion of placements and recruitments. Different students have different needs. Some may have the money but lack focus while others may have to work for some time to earn the money to start working on their decided goal. Many miscalculations occur due to lack of guidance and by imitating poor quality role models.

3) Special entry in top engineering colleges for school students who have scientific accomplishments: These are students with real aptitude and interest in science and technology but unfortunately they may not have the 'mugging up' skills required to score high marks in our examination system. These students are examples of 'meritorious but not topper' kind.

   Very often we read in the newspapers about school kids coming up with innovative ideas and solutions to our everyday problems. These children should be given special admission in their choice of engineering at top institutions, even if they are not academically performing. If we prevent their entry into scientific educational institutions, even after they prove their scientific talent and interest, just because they could not secure top marks, then they will most likely end up in a randcom arts college, losing faith in the system, where their talent and interest will be washed away over a period of time.

4) There is no national body to protect student rights from being violated by colleges. Just like how UGC and the education ministry are actively encouraging anti ragging campuses, to protect juniors from seniors, regulatory authorities should also protect students from college management that violate their rights and privacy. An organization like National Student Rights Commission should be set-up to take action against colleges and universities that mislead and violate the rights of students.

5) Engineering education should be flexible enough to support innovation and deviation from the prescribed path. Suppose an engineering student gets an idea like creating the next big social networking site or a new multi-port battery recharge device and if he/she wants to take a break from college after second year to work on this idea, that student should be allowed and encouraged to take an academic break with the option of returning after a gap of a couple of years and continue the engineering degree from where they left off. The university could also add an option of continuing through correspondence in case their idea flourishes and causes them to be too busy to attend regular classes.

It is sad to see that with the way the system currently operates, even if a student finds out how MS Excel can be improved further to create a better integration with MS Word and MS Powerpoint and if Microsoft wants him to join their company even if he has not finished his degree, the university will only let him go if he quits his degree course permanently. This is a risk that many parents won't encourage or allow. Some colleges instead of appreciating the idea will instead ask the student to pay the full tuition fee for the rest of the course and then get lost anywhere.

Let the student's degree certificate mention the gap in education and it will be up to the recruiters or future employers to decide that student's merit and not the university. Who knows how many Googles and Apples are buried and destroyed in the classrooms of our engineering cemeteries? We can find out if we allow these kinds of entrepreneurial breaks to students. Education has no purpose without an application. Such enterprising students add to the knowledge pool of the university and the college can actually benefit from their experiences.

Steve Jobs and Mark Zuckerberg had to discontinue their college to focus on their dreams and develop their ideas. In the Indian educational system that kind of approach is too risky, unless your parents are from an industrial powerhouse. We still need a basic under-graduate degree to even apply for a government exam or to apply for higher education. Allowing the students to leave the course in the middle with an option of rejoining later will add to their freedom and give wings to their creativity. This will also improve their risk taking ability. Very few Indians have created global brands and even the ones who did so were able to do it only outside India and mostly in western institutions of science and technology. If the colleges want their students to become like those inventors, they should first try to create an education system that allows as much freedom of thought, expression, innovation and support as theirs.

6) My degree, my elective: This sounds like the feminist concept of 'My body, my rights' and rights-wise, the application is same in the education system too. Students should be allowed to choose their own elective subjects. Some universities have interesting and useful electives related to climate change and renewable energy but these electives are not allotted to students because the colleges don't find it fruitful to employ specialized teachers for a single subject that many students may not even choose as their elective. Instead they assign the regular teachers to teach the elective, which are pre-decided for an entire class by the department. In a western engineering college, you will find one student studying ocean technology as elective and another student studying air defence as elective. Each student would have chosen an elective based on his/her interest. In India, you will find an entire class studying the same elective, chosen based on the convenience of the department.

If few students want to study an elective of their choice and are ready to face the academic consequence of poor exam performance, they should be allowed. Education is for the students, not the college.

7) All educational institutions should publish the financial details of the total fee collected, the amount they spend on salaries of the teaching and non-teaching staff and on the campus and building infrastructure. The difference will show how much money is pocketed by the college administration and whether the fee matches the quality of education that is given. Naturally the educational institutions will argue that this damages their privacy and administration. However, their arguments will also prove that their business structure is more important than the right of parents to know whether their children are getting their hard-earned money's worth.

   A popular educational institution in my city spent lakhs of rupees in building a place of worship within the college campus for the students and ironically this educational institution does not have any scholarship program for the financially unstable but meritorious students. A friend of mine who studied in the same religious-cum-engineering institute was warned that she will be prevented from attending class if she doesn't pay her semester fees. She was talented but poor. The best thing for the college to do is waiving off her fees and allowing her to study free of cost. Atleast the college could have given her a special consideration and given her more time. The college instead acted 'strict' and her father had to borrow money from his office colleagues to pay her fees.

This institution is a self-financing institution — it does not receive any government funding. The fee paid by the students is meant to improve the quality of their education and not for building a place of worship. It is sad that we have more religious places of worship than schools and colleges in this country. If it was the students' choice whether they should have an in-campus place of worship or better facilities in their labs and classrooms, I'm sure they would pick the thing that only their college can provide for them: improved educational facilities. Traditionally, we treat education on par with religion. Educational books are considered holy and divine and teachers are compared to the gods we worship. Ironically today many have converted both religion and education into a profitable and heartless business activity.

There are also other fancy institutions which spend crores of rupees on impressive pillars, arches and fountains but don't have any recent edition of engineering journals in the college library. There are also colleges that have sufficient water flowing in the fountain, to impress the parents, but not enough in the toilets used by students.

8) Engineering college students are still kids: Look at college students as teenagers, adolescents and

young adults but not as fully grown, matured adults. Students finish their school when they are 16 or 17 and they are usually 17 when they join college. For three years until the age of 20, they are still teenagers, with teenage hormones, teenage emotions and teenage dreams. They don't magically transform into adults just because they joined a college. In fact, for the major portion of their engineering college life, they are still teenagers.

There are some highly enthusiastic elders who think that just because they are 80 years old, they have 80 years of experience. They look at 19 year old college students and say, 'Look you are going to be 25 soon. Start behaving like a matured adult now itself.' A 19 year old college boy or girl can only behave like a 19 year old individual, with the mindset that is suitable for that age. When that person reaches the age of 25, then you ask him or her to behave like a 25 year old individual. This is exactly like how we force a 14 year old school kid to behave like a future IIT student.

We should stop stressing them out by saying things like 'look we are preparing you for the society, society will give you more stress in real life'. When students are made to feel stressed, depressed and rejected while they are still in college, they enter the society as stressed and rejected souls and contribute more to the social stress levels. There is a good chance that a stressed out school and college student will continue to be stressed for a long time in their adult life.

9) The students who want to study engineering but cannot afford to pay the fees of an engineering college or have not secured high marks in 10th or 12th board exams wind up joining poly-technic college. Poly-technic colleges are not the same as engineering colleges. You only get a diploma in engineering, not a degree.

   Poly-technic college students who study for a 'Diploma in Engineering' have the option of joining the second or third year of an engineering course. After school, these students join a poly-technic college and they can graduate as engineers if they manage to get lateral entry admission into an engineering college. Otherwise, they usually get a 'Diploma in Engineering' if they choose to graduate from a poly-technic college but unfortunately with a diploma alone they are not seen as 'graduates' by most recruiters. They are diploma holders and don't stand much chance against the engineering degree holders. These students would not have been able to join an engineering college directly after their school due to lack of money, marks and guidance but the lateral entry admission is a flexible advantage. Knowledge-wise there is not much difference between a poly-technic education in electrical science and an engineering education in the same electrical field. The word 'engineering' added to the word 'electrical' causes all the hype. It will shock engineering students when they see how easily a poly-technic student can rip open any motor and re-assemble it. What engineering students learn in theory, they learn in practice. Their knowledge level and talent is not, by far, inferior to engineering students just because they lack the brand value of an engineering degree.

Poly-technic students should be treated on par with engineering students and they should be able to get work after they finish their poly-technic diploma instead of having to waste their time with an additional academic year in an engineering college.

When I was studying electrical machines, it was these students who explained to some of our teachers how motors can be split and re-assembled. These teachers like engineering students have studied everything from a book, without even touching a motor in real life.

The employment market for those coming out of poly-technic college is very poor. They find work as lab assistants, electrical assistants and mechanical technicians under engineering graduates. Many even end up as electricians and mechanics. The irony is that someone who knows how motor works practically is made to work under someone who can only explain how it works in theory.

An electrician who knows more about the bulb, wiring and radios than an electrical engineer will barely earn 1/5th of what the electrical engineer earns. A mechanic would actually know more about a car and it's working than a mechanical engineering graduate. Companies that specialize in electrical and mechanical engineering work should open up their recruitment to poly-technic college campuses as well. Poly-technic diploma holders are looked up as the 'poorer cousins' of engineering degree holders. Many of these students only fall short in their communication skills, but as far as practical knowledge and focus are concerned they can match most engineering graduates in the country. What I say may not find favour among engineers but this is the hard truth.

Off late, campus recruitments are being extended to arts and science colleges. It is possible that in the future poly-technic colleges will also come under their recruitment radar. If it happens this will be one of the biggest challenges that engineering students will have to face to prove themselves in the campus recruitment market.

10) An IIT for every state: Practically this may not be possible in the next few years but if such an arrangement exists then ultimately students will benefit, competition will reduce and the hype that coaching-center advertisements give to IITs will end.

Brand-protectors of IIT will disagree saying it will dilute and affect the quality of the brand. What people like this fail to understand is that IITs don't exist for the sake of the name of IITs, they exist for students. Is the happiness of students who are citizens of this country less valuable than the brand value of an institution made of stones and bricks?

It is a selfish argument by many IITians to say that opening more IITs will bring down the quality of IIT. The problem with quality comes from the acute shortage of good teachers and lack of labs, not the availability of a better education for more people. Good salaries and investment in infrastructure will take care of the supposed problem of quality .

Thousands of children miss their chance at IIT by 0.1 to 5 percentage points. Does this mean that these students don't have the potential to be an IITian? It only means that the lack of seats and high competition decided the route the student is forced to take. When we waste our 'prestigious' IIT seats on future bankers and stock brokers, there is nothing wrong in opening more IIT's and creating more seats for students who have an actual interest in making engineering as their career.

Advice to Parents and Teachers

For Parents:

The first and most important question to be addressed by every parent is: What is important to you, as a parent? Would you rather have your children earn large sums of money every month or do you want your children to be happy with what they do in life?

Sufficient money can be made in almost every career choice. Don't consider children studying engineering as a matter of pride for your family. Your family's pride will shine more by saying that your son or daughter is happy with their careers instead of saying that they are in engineering or working in an I.T. company.

***

Your children don't owe you any respect just because you reproduced them. Respect is given on the basis of merit and on the basis of genuine love and care that you show towards your children and the freedom you give your children in allowing them to pursue their dreams. There are many parents who blackmail their children using the concept of 'respect'. 'Respect me and join engineering', 'If you don't get into the top institution, you are disrespecting me', 'If you don't get placed in campus, you bring disrespect to the family'. This is not respect in anyway. Neither is this respectful nor respectable. This is a form of blackmailing tactic used by parents to impose their ambition on their children.

***

What might have been extra-ordinary when you were studying has now become normal. Parents expect their children to understand them but how many parents actually make a conscious effort to understand their children? The best way to understand your child is to be their best friend. You are a great parent if your son or daughter could share with you the experience of their 1st drink, 1st kiss or at least their 1st arrear. You don't have to agree with or encourage them; you just need to listen without constantly judging and reprimanding. An unreasonable education and career target set by parents for their children had turned what could be a beautiful parent-child relationship into a hide-and-seek game. Be your child's friend and stand by them.

***

Children treat you the way you treat them. Forcing children to sacrifice their dreams to study for a college degree against their interest can be very counterproductive. Let's assume that it is done with the child's best interests at heart. Even then, the results of such an upbringing where the child does not have the freedom to choose his/her career path or dreams can be drastic: It can range from apathy to anger and in extreme cases even suicide. What is the point of crying later when your frustrated child feels rejected, leaves you or the world? Things have changed. Authoritarian or Hitler-type parenting had a place and time. But now it can't work. If you attempt it now, it will only result in a complete breakdown of communication and eventually the relationship will sour. The media makes a lot of noise when children abandon old parents but no one wants to go back into the past of these children and cross-check how their parents raised them. When you look at the number of parents who use their children as a dumping ground for their ambition, is it any surprise to see old-age homes getting filled with abandoned parents. This works like a parental karma. You admit your children in an engineering college against their interest and when they grow up, they put you in an old-age home against your interest. You get what you give.

***

Let us examine ourselves as parents. Are we seeking a return on investment for raising a child for 17 years by making sure they get into a lucrative career? Educating children has become a business but at least let's ensure raising children doesn't become a business investment.

You may think that 'forcing' your children to join an engineering or medical college against their interest, is being done for the child's future interest but you are only imposing your ambition on them. It is your interest to see them become an engineer and you disguise it as child's interest, to falsely satisfy yourself.

Don't discourage your son or daughter if they wish to deviate from engineering and join a Masters degree in a non-engineering course or start their own entrepreneurial venture. If you don't know what they are into, educate yourself about their future plans and encourage them. Before you ask your children to spend four years studying engineering, it's only fair that you spend at least four days to understand where their real interests are and what they want to do in life. Stop planning what your children will earn, where they will live and what car they will drive. Start dreaming about what sort of people you want them to be—upright, happy and valuable members of our society or frustrated and rejected souls hiding from everyone. Start thinking about how your children can be more happy even if they earn less. Trust me, you will not regret it.

For teachers:

There are good teachers and there are bad teachers. Good and bad are not defined by the quality of your teaching, which is also important but primarily your quality as a teacher is defined by the care you show towards your students.

Your classroom will be a diverse mix of all genders with different orientations towards life. Every student is unique. As teachers, you must understand conformity is different from uniformity. Every student is expected to follow the rules of the college but he/she is also entitled to live their college life in their own way.

***

If we observe those who fail to understand the idea of India, most of them are fundamentalist fanatics who look for uniformity rather than diversity. They either want all Indians to belong to one race and religion or assume that all Indians belong to one religion. I find a similar assumptive mentality among many teachers. They expect all students to think and behave in the same way like a robot. Every Indian is unique and every student is unique too. (In fact, everyone in the world is unique, before they are made to lose their originality to the crowd.)

Don't look for uniformity among students and don't impose uniformity on them. Just like how each teacher has his/her own style of teaching the same subject, students too have different studying habits and interests that vary from subject to subject. Every student cannot perform, behave and respond like the class topper. Each student comes to class with his/her own set of problems, confusions and dreams.

***

What students do in their personal life is their own choice. You should guide them if they indulge in anything unethical but don't use any moral policing tactics. Also ensure that what you consider unethical is not defined by outdated considerations of social morality. And you can't do this on your whims and fancies. If you do decide to take the moral high ground, you will have to do it regardless of who is engaging in wrong doings. For example, if the college management does anything unethical to harm the students, it is your responsibility to stand by the students and ensure that the interests of the students are not affected.

***

Many teachers are frustrated because they themselves are the same product of the same education and social system that is shaping or misshaping their students. The teachers of our teachers would have acted as poor role models and frustrated their school and college life. Now after graduating and becoming teachers, they imitate the role models they have seen in their college and school life. Unhappy and frustrated teachers produce unhappy and frustrated students, who after becoming teachers frustrate more students. Life is a circle. This is just like how sons and daughters learn 'anger' from their fathers and mothers.

***

When a student fails in your subject, don't label him/her as a poor performer. First find out whether that student has any interest in the subject at all. As long as the student studies without any interest, his/her performance will also be poor. Make them understand that passing the degree is important because that allows them to pursue their dreams once the course is over.

***

Just because they are students, it does not mean their life revolves only around the academic life of college alone. Like you, they too have a life after the college hours and outside their college life. Let them get on with it. Don't expect them to put everything else on hold for the sake of the classes and the course. I have known of colleges and teachers who expect the students to turn up for classes even if the student is sick or there is an emergency situation at home. Worse, they actually boast about this. Ironically, the social problems of teachers and students are not very different. Just like how teachers are worried about their marriage, spouse and children, students too worry about their relationship problems, family life and parental pressure. Just because you have graduated from college and became a teacher, don't fool yourself by assuming that students should not have any thoughts on love and relationships and should focus only on studies. The sooner you understand the reality of the student community, the better you will be able to help them deal with life.

***

When you sit at your table, while preparing notes for your class lecture, did you ever prepare them without your mind deviating into random memories of the past or worrying about your present family and social problems? Didn't you struggle to bring your focus to the classroom work at hand? You would have faced these mental distractions many times. If as a teacher, you face these distractions, imagine how distracting it will be for students. They too face the same social and emotional distractions, while preparing their assignments and studying for their exams.

***

When a student sits in the class not focusing on your teaching and has a distracted or confused look on his/her face, the last thing you should be worried about is 'your class lecture being wasted'. Obviously your classroom lecture does not matter to that student at that point of time. Either try to understand the student's problem in a gentle way or leave them alone. How would you feel if the Principal insults you in front of other teachers and students? Don't humiliate or mock them in front of the class. Students are not your punching bag to vent out your suppressed emotions.

Engineers and Non-engineers in my Life

These are people known to me, both professionally and personally. I have met them and come to know them during different phases of my life, from school to college to employment. Like all of us, they are also imperfect and were forced to take many bad decisions in all phases of life but I have learnt much from their successes and failures.

My Life: Confessions of a confused engineer

I never wanted to be an engineer and I have no idea what those two letters 'B.E.' are doing next to my name. As a matter of principle, I never write my name as 'Prabhu B.E.' When I was in my tenth standard, my understanding of education can be summed up in a single sentence: I need to get good marks in the board exam, so I can study science in a better school. My 11th standard was my age of 'losing innocence'. I see myself as an adult from that age period. The best thing about 11th standard in liberal schools is that they give you tremendous freedom because the students have to study a lot in the following year—the dreaded 12th standard. During my 11th, I started reading a lot of books and got opportunities to travel around the city for various school cultural and sports events. Reading a variety of books and meeting many new people opened my mind to various views of life and my mind started to expand. I suddenly wanted to try everything in life.

National Geographic magazines and Sidney Sheldon novels were my two windows to the world. I wanted a life that matched the images in the national geographic magazine and the text in Sidney Sheldon's novels. Few events in my life matched what I read in books but I dreamed of travelling around the world and trying my hand at various things.

After I finished my 11th standard, my parents admitted me into a tuition centre because they felt that the teaching in school was not enough for me to secure a seat in an engineering college. If my total marks could increase by even one more mark, it could make the difference between a good college and a bad college and the 10,000 rupees spent on extra coaching will have been worth it. The way my father spoke about engineering degree made me look at it as a passport to a better life. I had a difficult relationship with my parents. Hence, I hoped for a college with a hostel so I could spread my wings, leave the family nest and live alone. I did study well. I was an academically bright student. Of course, that does not mean I am a budding Einstein or scientifically intelligent. I just had the skill to memorize hundreds of pages of ten different books and reproduce them in sheets of paper. I cleared my 12th with more-than-good marks and secured an engineering seat in one of the top five engineering colleges in my state.

During my 12th standard and after my 12th standard board exams, I had very little understanding of the career and course options that were available to someone who is finishing school. My father wanted me to either join engineering or medicine.

Somehow those around me implanted a self-doubting thought about me being a doctor by referring to my poor drawing skills and I started seeing myself as a poor doctor material and decided that engineering is the better of the two, at that point in time.

One of my friend's father was in the state civil service and after meeting him in his office and reading about how civil servants are transforming the society, I suddenly had an interest in preparing for the Indian Civil Service exam. When I told my father about this interest and asked whether I could study law, he shouted at me saying that I wake up at 8 in the morning and to clear civil service I needed to study for 18 hours a day. (Years later I managed to clear almost all the government service exams, except the civil service, by studying for less than an hour in a day.) My father's angry outburst ended my interest in law. The only other thing that interested me was working for the National Geographic magazine but I had no clue about what I should study to work for them. If I asked my father, I was sure that he would burn all the national geographic magazines I collected and beat me more. So, I decided that I had no other option than to study what my father wanted me to study, which was engineering.

The next four years in an engineering college were a complete disaster. The first year was academically easy because I studied what I had studied already in my school and I scored above 80% but from the second year onward I slowly started failing in internal exams and finally in semester exams. Fortunately I cleared them in re-evaluation. My school friends were shocked. In school tests, I always scored above 195 for 200. My name was on the toppers list in both my school and tuition tests. Less than two years later, in an internal assessment test held in the 3rd semester of engineering, I distinguished myself by scoring a perfect zero! Bad as it was, it was only the beginning and the worst was yet to come. It was during this time that I realized that engineering is just not my cup of tea and that I could no longer push myself to wake-up early and memorize every page like I did in school. I wrote the few answers that I knew and most of the time I sat with a blank face in the exam hall. There was a government school next to my college, where children played. I looked at them every day and asked myself how things could have gone so wrong.

It is easy to be a poor performer in an average institution but difficult to be even an average performer in a top institution. Academically my school and college life were completely different. From being one of the best students in school and a school topper, I went to being one of the regulars in the college arrears list.

Though my marks were low, my understanding of the engineering subjects was as good as most studious students. Years after graduation, I still think I could have performed well if I was able to add interest to my understanding. To escape the burden of engineering studies and to satisfy my curiosity, I started taking part in lot of non-engineering activities outside college. By the end of my third year, I had tried everything from organizing blood donation camps to participating in Right-to-Education campaigns in IIT. That is the closest I ever got to an IIT, other than attending all their cultural shows, even if I had to skip my own college's cultural events.

Fortunately, my 70 - 75% marks was enough to appear in the campus placement and I got placed in TCS, a well-known corporate giant. I studied electrical engineering but usually only I.T. companies came for recruitment and so I had to learn computer science concepts to get into them. The sad irony is that though I got placed—with my poor percentage and not-a-few arrears cleared through re-evaluation—many of my college mates who were toppers could not get placed. The student who scored 90% got placed in the same I.T. company along with the students who scored 70%. Though their performance in the examinations were miles apart they were offered the same pay package by the recruiters. What was more cruel was that those who studied I.T. and C.S. could not get into I.T. companies but students who studied electrical and civil engineering managed to do so. Nothing exposes the hypocrisy of our engineering education system so well as that great melting pot called campus placement, where engineering students from all branches merge into the I.T. stream.

The placements were done by the 6th semester and at the age of 20, when I had barely understood the difference between an I.T. company and a Core company, I already had a job. Then, Tata Jagriti Yatra, an annual all-India train journey focusing on economic and social entrepreneurship called for applications. I came to know about it during my 7th semester exams, just a few days before the deadline for application submission. I knew I had to make a choice. I decided to close my books and started preparing for the short-essay questions of the Yatra. I failed in a semester paper but was selected to take part in a journey that became a life-changing experience. During the three-week train journey, I got the opportunity to write their pledge which we took during the new year's celebration. It was the first proper recognition for something I wrote. Throughout the final year in my school, I missed all writing competitions to study for the 12th board exams, to get into an engineering college. Now, in the final year of my engineering degree, I decided to even fail in a paper and put my campus placement at risk to achieve something through my writing.

***

There was a waiting period of 1 year after the final year of college, for the companies to issue their call letter. During the one-year waiting period, it became abundantly clear to me that this was not what I really wanted to do in life. Though when I look back, I can see I was not very clear about the direction in which I was going. After graduation, I decided I will try everything I wanted to do in life —from attempting civil service examination to appearing for the Army to starting my own global news analysis website. Two of my close interests after finishing college were photography and writing. Both found social media success. There were lot of likes, shares and comments on my digitally-edited photo-art and articles on current events. Some even went viral. But I had one problem which prevented me from converting even either of them into a full-time career. I was unable convert my ideas into a viable financial model.

After two years of self-imposed unemployment in pursuit of my artistic and social dreams, my parents decided that it is time I started working. I also started feeling guilty that I am being a financial burden and liability for my parents, even though it wasn't my fault that I ended up with a degree from which I could not benefit.

While I started looking for work, two years after graduation, I found that an engineering degree has no market value even if you have studied in one of the best colleges in the state. No I.T company was interested in hiring an inexperienced engineering graduate and there was no way I could get any experience if all companies kept rejecting me, by saying the same thing. Even the BPOs rejected me. Finally after six months of struggle, I found an Australian-time night shift job at a SAP technology company, through a good friend who referred me. I was paid less than what I was offered when I was placed in campus recruitment and my class-mates then were earning three-times my starting salary. Still, the job gave a sign of hope after many dark days. I met many non-engineers who were heading teams of engineers and those in their 40s who wanted to start their own business.

Most of the time I was among the few employees who were allotted night shift. Being alone and away from the family commotion gave me lot of calm thinking time. Soon I improved my focus and got selected for another job at one of India's leading auditing organizations. After my monthly financial income was secured, my mind came back to what I wanted to do since college. I decided finally to try my hand at writing a book. Even though I couldn't convert my life-interests to money, I didn't have much interest in financial wealth. What excited me above all was the chance of making a difference. I wanted to impact people, impact my society.

I realized that converting my ideas into a money-making enterprise may be rather tough. I believed I had artistic potential but I did not fool myself about its financial potential. My past record was not very encouraging: my first business venture at the age of 19 had failed because of a technical error and I could not run a business as a team. All my business and social ventures were one-man shows. When I experimented with photography, my digitally edited photo-art work had a good social media response. There were lot of likes, comments and shares but very few were ready to pay for something they could access on Facebook for free. This is similar to those fb prayer posts '1 like for 1 prayer' & '1 comment for god to bless the cancer patient'. I could not convert my art into cash just like how fb prayers could not cure a cancer patient. Indians and foreigners dumped likes in hundreds on my photo-art but I couldn't even sell one of them. A few galleries wanted to exhibit my works but they were not interested in purchasing them. It was the same with writing. My writing was received well on Facebook and some went viral on the website I posted. But again, I found at an early stage itself that making money through writing is beyond my literary ability. The main problem was that I don't write to suit what people want to hear. I wanted to tell them what they needed to hear. I wanted to speak about what was wrong with our society, culture and tradition. Unfortunately, attacking blind beliefs is not yet a profit-making model in this country. I had to make a decision. Since I was not after profit anyway, I decided to work full-time to support myself and focus on my hobbies and interests during after-work hours. At least till I either establish myself as a good writer or start a financially self-sustaining business model.

What you are holding in your hands right now is my first book. It took me almost two years to get my thoughts and emotions into this form. Along the way, there have been moments of doubts, and anguish and worry. Sleepless nights and restless days. The process of conceiving a book, writing it and wrangling with publishers to get it printed is like starting a company or birthing a baby. And like having a baby, you become emotionally so invested in this 'baby' that the thought of losing it simply breaks your heart. At least in a company, you can sell it to someone else if it doesn't work out, but if a book flops, no one will take it even for free. That is why I had to work full-time and write part-time. My writing may or may not be good. I may or may not have the potential to be a successful writer. But I think it is better to face the risk of embarrassment that comes from failure instead of the comfort that comes from safety.

If the book has the impact I hope for, my engineering degree would have finally been of some use to me.

My brother:

My younger brother, Pradeep Kumar, founder of HellBound Bloggers, is one of the few people whose interests, dreams and passion match with their college education. He wanted to be in the field of computers and internet before he was even 15 years old. His school marks added more strength to his dreams. His computer exam paper marks were always above 80% though he secured a just-pass in other subjects. He joined computer science engineering with an interest in the subject and with a desire to make his mark in the world of internet. He was working on blogs and networking at an amateur level when he was a school kid and in the 2nd year of his college, he started preparation for a professional website that focuses on technology writing. By the end of his third year, he had failed in many papers but had succeeded in creating a technology news website that gave him a decent source of advertising income every month. When he finished his fourth year and came out of college, he had a handful of pending arrears and a monthly revenue that was three times more than what his classmates selected through campus placement could earn in their tech companies. He spent the next one year clearing his arrears and professionalizing the site. When he finally cleared all his papers, he was ranked among the top 50 semi-professional bloggers of India, at the age of 21.

We both studied in the same school and I was academically better than he was. I topped the class and he often failed in his class tests. Our teachers, principal, relatives and parents advised him to be like me and told him that 'if you don't study like your brother, you will struggle in life'. But now he earns three to four times more than I do. My brother is the best proof of how when life-interest matches the right education, it will motivate our youngsters to realize their full potential and be happy with their lives doing what they like and enjoying what they do.

When we were in school, teachers used to point their fingers at me and tell him, 'Pradeep, you should study like your brother or else you won't come up well in life!' Now he gives employment to ten writers under him and I tell myself, like him, I should have also studied something I like in college.

Rahul and Karthik:

Rahul studied engineering for two reasons, first because he didn't know what happens after school and second, like in everything else in his life, his father made the decision for him. Karthik studied B.Sc. computer science degree in the best Arts and Science college in his state. Both of them didn't have any plan to create a future for themselves in the respective fields of their degrees. Their sole interests since their school days were in Cinema, Culture and Fashion. Though their under-graduate degree differed, Karthik had an advantage over Rahul. Karthik used his computer science course to learn film technology and movie-editing. He used his technical degree to obtain as much benefit as he could for using it in his cinema-related work. Whereas Rahul had to go for a post-graduate degree to learn about his field of choice. After graduation, Rahul rejected his campus placement offer and secured admission for a post-graduate management degree in one of India's best fashion management schools.

While Rahul pursued his fashion degree, Karthik had established himself in the ground-level work of cinema. His elder brother had worked as an Assistant Director in a major movie and Karthik had worked in the editing team of popular regional channels. For all practical purposes, it seemed like their paths had diverged. But then, the two friends decided to launch their own fashion cloth brand using traditional culture as a theme. Rahul once again rejected the campus placement offer he got in his fashion management school and Karthik shifted from his full-time job to a part-time job. Rahul was wealthy enough to indulge his entrepreneurial ambitions. But Karthik couldn't make a similar choice because his family needed him. He required a regular source of income and decided to work part-time till the newly launched business venture became self-sustaining and generated profit.

Together they started working on designs and researched the cultural backgrounds. They had two initial setbacks. While Rahul studied professional fashion management and Karthik was an expert in photoshop, neither of them could draft a design on paper nor balance their accounts. They could not even file their taxes on their own, though Rahul studied financial management as a subject for a semester (In fact, the only good thing Rahul seemed to have got out of his MBA-like school was meeting the love of his life.) But they did not let these things deter them. They hired experts to do what they couldn't do. After a year of research and ground work, they launched their fashion concept and it found instant favour and support among many shop owners and customers. Though the sales took time to grow and they are yet to cross the break-even point, reviews about their clothes in newspapers are encouraging and the duo are all charged up about the venture's future.

Prem Anand:

Prem took up computer science engineering because he had an interest in the field of computer science. Though his marks varied between average and good, he was brilliant as far as knowledge is concerned. But he got interested in the civil services during the course of his college study and was unable to decide whether to prepare full-time for the exam or work for two years and then start to prepare full-time. He decided to take a safer route and wrote the exam three years after his graduation. He was well-aware that his coding skills were top notch and with two years' experience, he can get back into the I.T industry later too. He was ready to focus on his real interest without any worrying thoughts about financial insecurity incase his attempt to crack one of India's toughest exam fails.

While preparing for civil service exams, he decided that even if he fails to become a civil servant, he couldn't go back to his office cabin work, after studying the country's problems in depth. Initially, he wanted to do a master's degree in computer science, if he didn't clear the exam and become another Steve Jobs. Now he has decided to apply for a law degree and take up NGO activism work. His new role model: Arvind Kejriwal, who successfully shifted from mechanical engineering to civil service to socio-political activism.

Karan:

Karan could not study engineering even though he wanted to because his parents could not afford the huge management fee and he didn't want to study in a poor quality engineering college. He had good marks and his parents admitted him in a decent Arts and Science college within the city. Though he was initially disappointed, he later picked himself up and decided to focus on his B.com degree with the hope that he would find some good employment with higher marks. His parents encouraged his artistic interests which included singing and playing drums. He participated in inter-school competitions whenever he could but neither he or his parents had any idea about the career prospects of studying music as a college degree course. He focused on his music during the weekends and participated in college cultural competitions. Eventually, he started working as a trainee under a drums teacher and played with him in music shows in the weekends. He did not earn lot of money but gained enough experience to play on his own without his teacher's guidance in a very short period of time. After graduation, all of us hoped that he will take up his music interest as a full-time profession. However, he joined a BPO to work on weekdays and play drums on weekends. Initially he said that a full-time music career is financially risky and that he would quit his job once he establishes his name in the field. Later I came to know that it was because of his girlfriend, who wanted him to have a steady job, instead of a random source of income. He does regret not pursuing a full-time career in music. While others I know destroyed their lives by listening to parents and relatives, Karan wasted his life-interest by listening to the love of his life. Unfortunately she did not realize that he had the same love which he had for her, for music too.

Gaganpreet Singh:

Gagan's father owned a chain of hotels and restaurants. His interest after school was to expand his father's hotel business. He had many ideas on how to do it even though he was unclear about what to study in college for his under-graduation. His wealthy father—who knew his son's long-term interest all along—could have let him study hotel management. There were lot of courses in India related to catering, food, hospitality management and business studies. With their kind of money, he could have even gone abroad to study food business but his father felt that studying a course about food technology was below their status and decided that he should study engineering. Fortunately, his father let him choose the branch of engineering course. He chose aeronautical engineering because he thought it was 'cool', when compared to other branches of engineering. He failed in some papers but managed to clear them with the help of private tuitions. Even before joining the course, his father told him that he can join his hotel business once he graduates and gets an MBA degree. He did keep his word. After his MBA, Gagan was put in charge of one of their small hotels so he could learn the ropes.

He spent 6 years for getting his Engineering and MBA degrees. If his father had let him pursue hotel management after school, he would have studied what he wanted to do in life and would have also saved six years of his life, which he could have used in learning about hotels in the field. Instead the only useful lesson he learned from his engineering degree was learning how students use the college canteen.

Mohit Gupta:

Mohit found it difficult even to clear his tenth standard board exams. He failed in physics and he re-wrote the exam, because his father wanted him to at least have a high school certificate. He knew that his son was not bright enough to get a college education and so he started coaching his son to enter into the family's pawn-brokerage business. Since he already made up his mind that his son should take over the family's broker business, he felt it made sense to help his son learn the rules and practices of the business as early as possible. Mohit is neither happy or sad about it. He knew that this was his best shot at earning money.

Sometimes he wonders whether if he had studied well, he could have also worked in an I.T company. He has his own shop now and earns about 50,000 per month, opens and closes his shop whenever he wants and lives a life of physical comfort, except for moments of occasional regrets. He earns more than what many engineers earn in the first five to ten years of their career. Other than the fact that he couldn't clear his tenth, I don't see any difference between his earning capacity and that of any engineering graduate like me. I am not saying there is no value to the degree because we earn less. But if money is all you are after, you don't need an engineering degree for that; you can do it with just your high school certification.

Sanobar Ahmed:

In the engineering college I studied, we had an Arts and Science college in the same campus, where a tuition friend of mine pursued her B.Sc. in microbiology. She joined the course because her parents thought that she would not find employment with an engineering degree. They decided that a science degree was enough for her and the money can instead be used for her younger brother's engineering college fees. Her academic life was not stressful. She had a great college life. At the end of the final year, she attended an off-campus interview and got selected in CTS. On the one hand, I am happy that major multinational Indian I.T companies are slowly recruiting Arts and Science students for posts supposedly intended for only engineering graduates. On the other hand, I am surprised at the economics behind this. A college-mate who finished engineering along with me also got placed at the same company. He spent around 2.5 lakhs for his four year engineering education and she spent less than 60,000 for her 3 year science education. But ultimately she achieved what we did at $1/5^{th}$ of the total cost.

For many engineering students from all streams, the immediate goal before finishing the college degree is getting placed in an I.T company. They spend a minimum of 2 to 3 lakhs to achieve this. It goes up to even ten times the minimum amount for some. As recruiting companies spread their net to include more and more Arts and Science colleges, it should make engineering graduates wonder whether the campus placement they achieved could have been achieved at a fraction of the cost and energy they wasted in an engineering college. But students like Sano should be proud of their achievement, even if they don't seek a career in I.T.

Priyanka Kumar:

My friend's mother belongs to the 1960s to 1980s era, where people go to college, just to study in a college and not for a particular course. The focus then for middle-class and lower-middle class families was studying in any college and in any course. She studied marathi Literature because it was the least expensive course available in the college closest to her house but she later developed an interest in the subject over a period of time. She then worked as a teacher in a government school. Unfortunately she could not continue working after she got married. She resigned from her job in 1984 to take care of her children, even when she was earning more than her husband. It's been almost three decades and not much has changed when it comes to women resigning after marriage. While we can dismiss this as a trend of pre-2000 India and not present in 21st century India, the reality is that many female engineering graduates from leading institutes still quit work after marriage.

Ganesh Rao:

For Ganesh's lower middle-class parents, their son joining an engineering college was an exit route from years of financial struggle. Ganesh secured an education loan for 8 lakhs and at the end of his course, he got placed in an I.T company. When he was in his 3rd year itself, his father retired and it placed additional stress on him. Somehow he managed to keep his percentage above

85 and got placed. Now his problems didn't end after he started working. He had to give a portion of his salary to his parents and use the remaining amount to pay the monthly dues for the loan. He saves as much as possible, does not go out with his colleagues to parties and even avoided asking the girl he liked for a date. Soon both his emotions and health took a hit. Three years after his graduation, he is yet to pay off his dues and debt fully. He still looks at joining engineering college and working in an I.T company simply as a duty to his parents. But what about his life now?

Priya and Rekha:

Priya and Rekha are cousins who are of the same age. Rekha wanted to be a civil engineer while Priya wanted to study in an arts college. Ambition-wise both had similar views on college and career but the problem was that Rekha was academically poor and scored less while Priya performed well in the 12th board exams.

Their parents decided that Rekha should join an Arts college as a punishment for not studying well in school, even though her rich father could have gotten her admission through management quota and Priya had to join an engineering college as they felt that Arts colleges were status-wise lower for her good marks. Both ended up studying what the other wanted to study.

Popular Deviants: You are not alone

Many famous engineering and science graduates have deviated from engineering and found their real passion and life interest in a non-engineering field and have made it big in life. If you also want to shift to a non-engineering side after spending four years in an engineering college, don't be sad, disheartened or embarrassed.

Though you may not agree with the political or social views of some individuals listed here, use the list to look at their 'shift'; look at how courageously and successfully they made the transition from engineering to a non-engineering one. There are enough success stories in India to show us the way. What matters is finding your life interest and learning how to make a career out of it.

1) Raghuram Rajan and Duvvuri Subbarao: both IITians who pursued economics and served as governors of the Reserve Bank of India.

2) Manohar Parrikar: metallurgical engineer from IIT Bombay, who later joined politics to become the Chief Minister of Goa.

3) Arvind Kejriwal: a mechanical engineer who joined the Indian Revenue Service and then quit it to start an anti-corruption movement which lead to the establishment of his own political party. He then became the Chief Minister of Delhi.

4) K. V. Kamath: a mechanical engineer who pursued management studies at an Indian Institute of Management. He went on to become the Managing Director and CEO of ICICI bank.

5) Hemant Karkare: a mechanical engineer who joined the Indian Police Service and became the chief of Mumbai Anti-Terrorist Squad. He was killed in service during the 26/11 terrorist attacks.

6) Rowan Atkinson: better known to the world as Mr.Bean: graduated with a M.Sc degree in Electrical Engineering. He pursued his passion of acting, and went on to become one of the most successful actors/ comedians in the world.

7) Jairam Ramesh: another Mechanical Engineer who shifted his academic focus to pursue Public Policy and Management at Carnegie Mellon University. He has served in the Planning Commission, Ministry of Industry and other economic departments. He was later elected as a Member of Parliament and held ministerial positions in Environment and Forests, Rural Development, Drinking Water and Sanitation.

8) Deep Joshi: a mechanical engineer who moved into the social service space. He started PRADAN,

an NGO that promotes rural livelihoods and helps around 200,000 families, more than half of them from tribal areas.

9) Nitish Kumar: studied electrical engineering from Bihar College of Engineering but turned to politics and became the Chief Minister of Bihar.

10) Tom Scholz: did his bachelor's and master's degrees in mechanical engineering from MIT. He invented a guitar amplifier and worked for the Polaroid Corporation as a product design engineer. But he is best known as the founder of the rock band 'Boston'.

11) Brian May: graduated with a B.Sc. Degree but became the guitarist of British rock band Queen, to rock the world by composing 'We will rock you'.

12) Alfred Hitchcock: studied engineering at the London County School of Engineering and Navigation before entering the film industry and starting a revolution in the world of cinema.

13) Dolph Lundgren: a Swedish chemical engineer who acted as Ivan Drago in Rocky. He is a successful actor, director and martial artist.

14) Our own home-grown Actor Madhavan graduated in electronics engineering.

15) Tom Landry: One of the most famous football coaches in the NFL history has an engineering degree.

16) Ashton Kutcher: dropped out of biochemical engineering to pursue acting and become a heartthrob of millions.

17) Verghese kurien: a mechanical engineer and the brain behind Amul brand is called the father of the white revolution for initiating one of the world's biggest agricultural development programs.

18) Vikram Pandit: Citigroup's youngest CEO holds a Bachelor and Master's degree in Electrical Engineering. He later finished a post-graduate course in finance.

19) Chetan Bhagat: probably best known for bringing to national attention, the flip side of the IITs and the engineering success stories through his writing, graduated from both an IIT and an IIM, and later left his banking career, to take up writing full-time.

20) Bedabrata Pain: a scientist who had earlier worked with NASA and was member of the team that invented CMOS image sensor. He recently won a National Film Award for Best First Film of a Director.

21) Cindy Crawford: dropped out of chemical engineering course to pursue modeling and acting.

22) Kailash Satyarthi: Our very own Nobel Peace Prize winner was an electrical engineering graduate, who left his engineering career to work for child rights and liberated over 80,000 child labourers.

That is a very small list but there are many, many more such real-life heroes and heroines.

***

Many successful people have quit their engineering career to focus on what makes them happy and have triumphed. For many of these people, when they attempted to leave their engineering life, people around them would have mocked them but now after they succeeded, they are praised.

That's the way things are and will be. Whether it is engineering or anything else, every time someone leaves the beaten path to pursue something new, there will be criticism and snide remarks. The doomsday pundits will threaten you. When you leave your engineering degree and become a movie director, your relatives will laugh at you. But when you receive an Oscar, they will wait for your autograph and beg you for a chance to feature their son in your next movie. If you choose to start a political party or a social activism group after leaving an engineering career, cynics will mock you. But when you finally make your mark, they will stand up when you enter. This is how people are.

So, don't worry about what others think about your non-engineering plans. If you succeed, it is good. Even if your plans fail, doing what makes us happy is never considered a failure.

Is the Trend Changing Now?

The good news is yes. It is now fairly common to see fresh engineering graduates — as well as engineers with years of experience in the I.T and engineering sector — quit their field of study to pursue their field of interest in various non-engineering sectors including Art, Media or Social Service. Many engineers have also become successful full-time writers, photographers, fashion designers and actors.

Perhaps it's time to have a National Wasted Engineers day on April 1st, to celebrate engineering graduates who branch out into non-engineering fields. While this changing trend is heartwarming and encouraging, it will be way better if students directly join their favourite non-engineering field right after school, instead of spending four years in getting an engineering degree and then spending four more years trying to figure out what they want to do in life. Even a doctor studies MBBS for 6 years but engineers spend four years studying engineering and then waste many years wondering why they studied it.

***

Many recent events have contributed to this change. Chetan Bhagat's book, Five Point Someone, which focuses on the frustration of students forced into engineering and it's successful onscreen adaptation by Aamir Khan in '3 idiots' and Vijay in 'Nanban' sent out a hard hitting message that 'All is not well'. The book and the movies show quite poignantly how students are tortured and brain-washed to become engineers, losing their individuality and independence in the process. We are made to study engineering with a false promise of happiness. Only those who graduated know that there is no happiness in studying something you don't like and no happiness is ever found this way.

***

The focus of today's youth seems to be about enjoying life, which is good without any doubt. But this 'life' which they seem to be enjoying is only restricted to the after-work time in the evening and the eternally-longed for week-ends. The idea that you could enjoy work as a part of life doesn't even occur to most because almost all are trapped in the wrong jobs. In contemporary India, there is rarely a focus on the 'present'. The entire focus of our life philosophy is oriented towards the 'future' or the next stage. When we are born, we are prepared for school. When are in school, we are prepared for college. When we are in college, we are groomed for work. When we take up work, after 21 years of education, we sit and wonder whether we could have enjoyed school and college life differently.

Despite the glamour of six-digit salaries and posh office cabins, many parents and society in general, have no idea about the frustration and stress that comes with an engineering and I.T. work role. All engineering positions from software coding to automobile engineering require continuous up-gradation of knowledge. You need to update yourself continuously about the latest coding platforms, mechanical advancements and new inventions all your life. For many of us studying these as a pre-fixed syllabus itself was an academic nightmare. Imagine doing this on a life-long basis! We study engineering as an opportunity for a better life, falsely telling ourselves that with an engineering degree, we will lead a stress-free life. But this is far from the truth. Imagine the shock and disappointment, when we discover we not only have to live every day with the frustration resulting from the gap between our area of work and our area of interest but also deal with the chilling reality that this promise of so called stress-free career was all along a lie. The gap between our life and life-interest is responsible for many of our sad emotions and frustrations.

***

'What do you want to do in life?' is a question that teachers, friends, parents and even random strangers often throw at kids. This question so casually asked sometimes just to break the ice, scares the hell out of most of us. We try to manufacture an answer to satisfy the question and more importantly the questioner, and often end up giving a response that the questioner finds 'cool' and 'normal'. If we don't have a cool answer, we are looked down. We keep looking for approval from others and many to just answer the question generate future plans like I.T. career, engineering job, M.Tech, M.B.A, etc. If you don't have a plan, admit it and just tell them that you are yet to arrive at a decision. Even if you give a 'normal' answer to get away from the question, don't take it as an answer for yourself.

***

These are some things you can do, to help yourself.

Seek professional help:

Do you feel confused and cornered? Don't worry, you are not alone. Most of us are sailing in the same boat. We live in a country where people pay 5 lakhs to 10 lakhs to rent out a wedding venue, 2 lakhs for the photographer and videography, and another 5 lakhs for the food and hotel stay; and yet, when the marriage comes under strain or breaks down, they shy away from paying 5000 rupees to a marriage counselor and seek professional advice. This is how we are. Traditionally, we Indians avoid seeking help for any life problems. We either think that God will take care of our problems or good things will happen in the end. I am sure God has better worldly problems like starving African children to take care of instead of sitting and planning your career move. And the problem with the repeated saying 'good things will happen in the end' is that, your life will be over when that end comes.

Your life is in your hands. You are its driver. Take control of it and guide it to a better road. There is nothing wrong in seeking professional advice for your personal problems, be it in your career, work, relationship, education or life.

Take someone into confidence:

All of us need some sort of emotional counseling throughout our lives. If you have a good real-life friend/ cousin or even a facebook friend who is able to listen to you and guide you, then that person can act as a emotional counselor. But the problem is many youngsters don't have anyone to talk with openly, even if they always have a gang of 'friends' around them. Often we hesitate to share our problems and end up bottling it all up. It is not a surprise then that India is rated among the world's most depressed nations. We don't open our mouth and say 'I need help.' We sit at home and cry but we don't try to find out why we are not able to find happiness. Sometimes when we talk to others about our problems, that itself can calm our mind and allow us to think clearly.

Introspect:

Many writers and self-help gurus ask youngsters to look within but we don't know how to do that. We don't know how to find ourselves amidst all these confusions and conflicts in our mind and heart. We spent so much time polishing our image for the society that we have now forgotten who we really are. Most youngsters face this conflict that professionals call identity crisis: The tension between the two identities – the identity which we have painstakingly crafted for others and our real identity — causes an emotional breakdown often. Once you separate the two, you will know the difference between the life you are living for yourself and the life you are living for others. If you manage that, you have made a good start in finding an escape from this conflict.

Experiment and Experience:

It is easier to understand what the world wants but not what you want in your own life. If you have not yet found your life-interest, there is no need to feel guilty. Our life-interests come from what we experience and how we interpret our experiences. That is why it is best to read a lot, meet lot of different people and travel a lot when you are young. When your mind explores the outside world more, it helps you understand yourself better. You need not pack your bags and travel all over India and the world. You need not stuff your room with books. It is enough if you even visit different parts of your city every weekend for a start. You need not become a bookworm, but read. And read widely. Try to read across genres like fiction, non-fiction, mythology, current affairs, national geographic magazines, etc. When your mind spreads more, you see lot of new things about life and see old things with a new perspective. A well-travelled and a well-read person always has a better grip over his/her life.

Shake free from social media control:

What many nowadays look for is social media appreciation in the form of Facebook likes and twitter retweets. If you post about a job selection in a fancy company, you will get hundreds of likes and comments even if you don't like the job. But the joy that comes from Facebook appreciation is short-lived. Once you get used to the job and once you realize that you are not enjoying your career, then even a thousand likes won't matter.

It is sad to see youngsters planning a tour, not to a place they can enjoy, but to a place that looks cool while posting on Facebook. Feel free to share and post what you like but don't live a life for Facebook attention and attraction. This is just like how many live a fake life for society's appreciation, instead of living their original life. Now don't go and de-activate your Facebook profile. It is good to be active on Facebook but don't get carried away by the 'like' mob. Truth is truth even if no one 'likes' it.

***

A word of caution: Many may not quit their careers even if they don't like for various genuine reasons like family pressure or survival needs. Let us tread gently here. While the commercial trend is to mock them and make them feel bad, ultimately this will not serve any purpose other than making them feel more sad, desperate and ashamed. We need to encourage and guide people trapped in wrong jobs, to pursue their life-interest and help them make the necessary changes to get the best out of their lives.

***

There is a growing recognition today that a happy low- paying job is better than a stressful high-paying job. If there is one thing, every engineering student or graduate can do to strengthen this recognition, it is this: put an end to children being forced to study what they don't like. Tell the people the truth. If we could destroy the hype that is being showered on engineering degrees, that alone will be the biggest investment we could ever make for our future generation.

***

Entrepreneurs—the new breed of uncles and aunties: It is very cool now for young celebrity entrepreneurs to ask youngsters to become entrepreneurs through their TED talks and fb pages. I don't see any difference between them and my parents. We should not tell others that they should 'be engineers or entrepreneurs' or 'be this or be that'. We can only ask them to do what makes them happy or help them find out what makes them happy. It is funny how some of the entrepreneurial youngsters of 21st century India who rebelled against the uncles and aunties of 20th century India have started to resemble them. The pressure to imitate and follow the crowd is dangerously high both in engineering as well as entrepreneurship. The challenge everywhere lies in maintaining your originality and standing up to the world that constantly tries to replace that originality with uniformity.

Remember what we said in the Introduction? To build a happy India is more important than to build a wealthy India. Don't follow the crowd just because they told you to. At the same time, don't try to do something different. Do something you like. Even if it is engineering! That is the best way to be happy.

Postscript

India's Project Centres: Proof that engineering graduates are not engineers

During the course of a four-year engineering degree most students are required to complete two projects – a mini-project and a final year project. Let's first look at how they actually should be. The mini-project, done usually in the second-year, consists of a minor scientific idea conceptualized and implemented by the students. Similarly, the final year project done in the last semester focuses mostly on a critical original scientific work formulated, researched and implemented by the students, on their own, with little or no guidance from the teachers.

This is how college projects should be done and this is how they are done in the scientifically advanced nations of the world. What do we do in India is another story though.

Step 1: Find the project centre closest to hostel/home.

Step 2: Ask the project centre owner (who usually calls himself as a director) for the current list of IEEE papers and the cost price for implementing the designs on circuits and machines.

Step 3: Decide the ideal project based on group budget. If there is a rich student in the group, you are damn lucky. You can now choose a high-tech robotic project. Else we will have to stick to the ordinary ones.

Step 4: We tell the project centre owner that we will 'consult' the teachers and return with the chosen paper. By now, you'd know 'Consult' does not mean a technical discussion with the HoD; it simply means taking a print-out of the IEEE paper and making a power-point presentation before the teachers. The teacher will go through the student project list and cross-check whether any other student has already purchased the same project before you. If any other batch has taken the same IEEE paper, then we have to go back to Step 1.

Step 5: Once you have zeroed in on your 'project', pay the project centre owner and update Facebook status as 'Yay! Finally we have decided on our project. We are going to be engineers soon!' If you cough up an additional five hundred to the lab technicians who create and design the project, they will explain the most fundamental concepts in your circuit better than your college teachers. Now all you have to do is appear before the college panel, as if the entire project is based on your own idea, on which you worked for years.

Step 6: Present it before the teachers and invigilators and secure a pass-mark.

Step 7: Congratulations, you are an engineer!

***

If University Grants Commission or AICTE creates a rule that students are banned from purchasing college projects and all projects should be based on their original thoughts and ideas, then I can assure you that over 90% of India's engineering graduates (including me) would have never finished our degrees. Because we have no idea what we studied and why we studied it.

Have you started thinking yet?

Acknowledgement

I would like to express my gratitude to everyone who helped me with the book; to all those who provided support and spent their valuable time in reading and giving ideas on further improving the themes originally conceptualized.

I thank my wonderful editor Tanya Francis, for shaping this book into a readable form and publishing consultants Prithvee, Surekha, Yamini, Nishanth M, Sakhi Ahamed, Subhhaiya Perumal and Anas Ahmed at Notion Press for being patient with my million doubts and for assisting me throughout the book publishing process. I am grateful to the help provided by Abanti Dutta in proofreading the chapters, despite her busy academic schedule.

Thanks to my brother, Pradeep, founder of HellBound Bloggers, for being brutally honest about my writing and for being the first critic. Thanks to Kumar and Raghav, founders of Angi Clothing, for helping me understand the transition from a frustrated engineer to a successful entrepreneur.

Thanks to Swathi Jain, for her help in connecting me with the right people and to Dipti Tambe for her wonderful hospitality at the JNU campus, which helped me understand how a student life should be and also made me regret not taking up arts, instead of engineering.

Thanks to my Gyansters cohort and fellow yatris at Tata Jagriti Yatra, for helping me understand the diversity of life and India.

I express special thanks to Arpita Tiwari, Srivatsan Raman, Diksha Sorate, Bhuvana Jayaraman, Aruna Ramachandran, Julian Rajiv Martyn - my facilitator at CRY-Child Rights and You, Nancy Thomas and Vidya Reddy of Tulir - Centre for Prevention and Healing of Child Sexual Abuse, Dr.Pragya Kafley, Irene McCullagh, Pooja Sharma, Tony Daniel and his team at Capgemini, BS Vishal of Viveks and Abinaya Selvam of Uptown Ideas.

I must mention this here more than anywhere else in the book. I am indebted to those who shared their experience and thoughts with me. I would also like to acknowledge the people who contributed to the bad experiences in my life. Without you, this book would not have been possible and I would have probably written a happy-college-life romance story.

About the Author

Prabhu studied electrical engineering in one of the leading engineering colleges in his state but like many of you he never pursued a career in engineering. He ran a news analysis website few years earlier while working for a technology consulting firm. He took part in the Jagriti Yatra, during his college days and now he works for a national auditing organization in a financial position.

His first job, however, was at a bookstore where he worked to pay for his arrear re-evaluation and to read books for free. He also holds a diploma in journalism and this is Prabhu's first book.

His other interests include photography and travelling. He lives in Chennai.

You can contact him at

facebook.com/prabhu.india

facebook.com/WastedEngineers